A NATURALIST'S GUIDE TO THE

BIRDS OF THE PHILIPPINES

Maia Tañedo, Adrian & Trinket Constantino and Robert Hutchinson

JOHN BEAUFOY PUBLISHING

Reprinted in 2024

This edition first published in the United Kingdom in 2018 by John Beaufoy Publishing Ltd
11 Blenheim Court, 316 Woodstock Road, Oxford OX2 7NS, England
www.johnbeaufoy.com

10 9 8 7 6 5 4

Photo Credits
All photos by **Robert Hutchinson** except as detailed below.
Front cover: *main image* Palawan Peacock Pheasant, *bottom left* Red-bellied Pitta, *bottom centre* Wattled Broadbill, *bottom right* Magnificent Sunbird.
Back cover: Palawan Flowerpecker.
Title page: Flaming Sunbird © Irene Dy. **Contents page:** Philippine Dwarf Kingfisher.
Main descriptions: photos are denoted by a page number followed by t (top), b (bottom), l (left) or r (right).
Mike Anton 63b, 89t. **Nilo Arribas Jr** 53b, 55t, 142b, 147b, 148b. **Adrian Constantino** 10, 13, 20b, 25b, 26b, 46t, 50b, 57b, 59t, 82tl, 88t, 93b, 109t, 128b, 129t, 139b, 145t, 152b. **Trinket Constantino** 11, 17b, 21b, 48b, 75t, 97bl, 127bl. **Rommel Cruz** 57t, 121t. **Bram Demeulemeester** 28t, 70b. **Irene Dy** 37t, 39t, 42t, 49b, 51tr, 58b, 65b, 100b, 115t, 131b, 146b, 149b, 151b. **Nicky Icarangai Jr** 150b. **Jops Josef** 61b, 84t, 134tl. **Bob D. Kaufman** 33b, 35t, 150t. **Ixi Mapua** 81t. **Tony Palliser** 60b. **Sylvia Ramos** 77b. **Tina Sarmiento Mallari** 86t. **Pete Simpson** 25t, 31b, 36t, 46b, 97t, 101t, 119t, 127br, 137t, 144b, 152t. **Maia Tañedo** 7, 14, 19b, 20t, 42b.

ISBN 978-1-912081-53-0

Edited by Krystyna Mayer
Designed by Gulmohur Press
Printed and bound in Malaysia by Times Offset (M) Sdn. Bhd.

·Contents·

Introduction 4

Geography, Island Groups and Major Weather Patterns 4

Avian Biogeography 5

Habitat Types and Bird Communities 6

Bird Migration and the East Asian Flyway 8

Conserving the Region's Birdlife 8

Birdwatching Opportunities in the Philippines 9

Nomenclature and Taxonomy 15

Species Accounts and Photographs 16

Checklist of the Birds of the Philippines 156

Further Information 171

Index 173

INTRODUCTION

The Philippine archipelago comprises more than 7,000 islands, the majority of which are uninhabited. They lie south of Taiwan, east of Borneo and north of Sulawesi, covering a land area not much bigger than the United Kingdom. Nevertheless, the island biogeography and diverse habitats have created a haven for evolution, and they contain an astonishingly rich biodiversity, including 190 mammal, 100 amphibian and 300 reptile species, and 3,500 trees.

High levels of endemism make the Philippines one of the world's biodiversity hotspots and one of the most biologically rich countries in the world. The high endemism level is well represented by the avifauna, with 683 species currently recorded, of which 239 (to date) are recognized as endemic to the Philippines.

Birdwatching in the Philippines is an exciting adventure that can take birdwatchers to different parts of the country to not only see the bird species unique to each area, but also experience the various landscapes and cultures each province has to offer. The various sites that can be visited are as diverse as the species that can be seen. Different habitats and landscapes offer a wide range of experiences, from urban areas and marshlands, to forested mountains and tropical islands.

This book provides basic information about the Philippines and gives an idea of the opportunities for birdwatching in the country. The 280 bird species featured, all illustrated with photographs, merely scratch the surface of what is available, but they provide an idea of what can be seen.

GEOGRAPHY, ISLAND GROUPS AND MAJOR WEATHER PATTERNS

The Philippines is an archipelago that lies in the western Pacific Ocean in Southeast Asia. It is bordered on all sides by bodies of water and has a coastline totalling 36,289km. To the north lies the Bashi Channel. The northernmost islands of Batanes and the Babuyan Islands lie on the Luzon Strait and face Taiwan. To the east is the Philippine Sea, facing the Pacific Ocean. The west is bordered by the West Philippine Sea and faces Vietnam, while the south is bordered by the Celebes Sea and faces Indonesia.

The Philippines' 7,107 islands have a total land area of 300,000sq km. The islands are grouped into three major island groups: Luzon, Visayas and Mindanao. Luzon is the largest island in the Philippines, and together with the islands of Marinduque, Masbate, Mindoro, Palawan and the Batanes Islands, makes up the Luzon island group in the north of the archipelago. The Visayas island group is located in the central Philippines and is made up of the islands of Bohol, Cebu, Negros and Panay, the largest island in the group. The Greater Mindanao group, located in the southern Philippines, includes the large island of Mindanao, the islands of Samar and Leyte, and the Sulu Archipelago, which is composed of Basilan, Sulu Island and Tawi-Tawi. The three major island groups are divided into 17 regions, which are further divided into 81 provinces.

The country sits on the Pacific Ring of Fire and its islands are volcanic in origin. Numerous volcanoes have been recorded throughout the Philippines, but only a small number are said to be active. These include the Mayon Volcano in Albay, also famous for its perfect cone, Mt Pinatubo in Zambales and Taal Volcano in Batangas. The Philippines also has many mountainous regions, a number of which are still covered with tropical rainforests. The highest mountain is Mt Apo, which has been declared a national park and protected area. It rises to 3,146m above sea level and is located on Mindanao.

The Philippines has a tropical climate characterized by high temperatures, high humidity levels and abundant rainfall at certain times of the year. The average temperature is 26.6°C, with the coolest months being January–February, and the hottest temperatures peaking in May. Generally, the climate can be divided into two seasons: the wet season and the dry season. The amount of rainfall varies from region to region, and is mostly affected by the seasonal prevailing winds. The south-west monsoon, locally called *Habagat*, is the wet or rainy season that falls in June–November. These months experience the most rainfall and hot, humid weather because of the winds coming from the south-west, as well as typhoons. Tropical cyclones usually form in the Pacific Ocean and make landfall along the eastern part of the country. The north-east monsoon, or *Amihan*, occurs in November–May. It brings cooler winds and little to no rainfall to many parts of the country. The dry season can be further divided into the cool dry season during December–February, and the hot dry season in March–May.

AVIAN BIOGEOGRAPHY

The archipelagic nature and tropical climate of the Philippines has led to an explosion of endemic species that is reflected in its avifauna. The unique geologic history of the islands has allowed the evolution of distinct species and subspecies (or races) often different from the rest of South and Southeast Asia, contributing to the high rate of endemism. Land bridges that formed during the last ice age connected many of the islands that are currently separated. These now represent the major faunal regions of Greater Luzon, Mindoro, West Visayas, Greater Mindanao, Greater Palawan and the Sulu Archipelago. Other islands have never been connected to any of these larger land masses, and have developed their unique flora and fauna: Batanes and Babuyan Islands in the north, Siquijor in the Visayas, the Tablas-Romblon-Sibuyan group west of Mindoro, and Camiguin Sur north of Mindanao.

Even after separation into the present-day islands, unique species still emerged. For example, Cebu Island, which is part of the West Visayas faunal region, is considered a distinct Endemic Bird Area with three endemic species and several endemic subspecies not found on nearby Panay, Negros or Masbate. Palawan, which largely originated from the Asian continent, has birds that show more similarity to birds in Borneo and the Sundaland than to species in the rest of the Philippines. On the larger islands of Luzon and Mindanao, extensive mountain ranges of various altitudinal ranges have allowed regional differences in avifauna, separating the islands into likewise unique, biogeographic subregions. For example, the island of Mindanao is further divided into the Eastern Mindanao, Liguasan and Zamboanga Peninsula subregions. In the faunal region of Luzon the avian diversity on the Cordilleras can differ greatly from that in the Sierra Madres, the former mountain range cutting across central northern Luzon and the latter across the eastern side of Luzon. The smaller Zambales mountains also separate the Zambales and Bataan provinces on the western side of Luzon facing the West Philippine Sea.

Much of the Philippines' biodiversity is still understudied and new species are still being described. The Calayan Rail was discovered in 2004 on an island of the Babuyan group, after which it was named. With further studies in morphology, vocalization and genetic characteristics, as well as now routinely used molecular techniques, many species have been split or even assigned to new taxonomic groups. Based on vocal divergence, for example, the Philippine Hawk-Owl complex has recently been split into seven species. Many more changes in the taxonomy of Philippine avifauna can be expected in the coming years, which will increase endemism even more, as the current rate of endemism in birds is thought to be grossly underestimated. Endemic species may, in fact, increase to almost 50 per cent, much higher than the currently recognized 35 per cent and far closer to the endemicity percentages for Philippine mammals and herps.

Habitat Types and Bird Communities

A variety of habitats can be found throughout the Philippines, from coastal mudflats and plains, to grassland and mossy forests. Different regions have different terrain and habitat types, which serve as homes to many plant and animal species. A specific region can have a number of these different habitats. Several mountainous regions can be found across the country, many of which have been declared protected areas.

FORESTS

The Philippines is lush with tropical forests. There are still patches of natural and old-growth forests with indigenous trees undisturbed by humans, but most forested areas are now secondary forests that have regrown after logging. Forested regions can be found in a variety of locations, from beaches to lowland areas. Aside from the tropical rainforests, the country also has different types of forest cover, including:

- **Mossy Forests** Montane forests in high elevations and mountainous regions.
- **Mangrove Forests** Clusters of different mangrove species growing along tidal mudflats and wetlands.
- **Beach Forests** Wooded areas comprising woodland growing along sea coasts and beaches.
- **Residual Forests** What has been left of forests after their use for commercial logging. Although generally disturbed by human encroachment, these forests serve as homes to a large percentage of the birds in the Philippines.

COASTAL MUDFLATS AND BEACHES

With more than 7,000 islands, the Philippines has abundant coastal habitats and beaches. These areas, particularly those with extensive mudflats, are popular wintering sites for migratory birds. Huge numbers of waders can be seen on exposed mudflats during low tide, foraging in mixed flocks. The Olango Island Wildlife Sanctuary in Cebu is a popular coastal site for migratory birds including Asian Dowitchers and Chinese Egrets. It is one of the six Philippines' RAMSAR sites declared as wetlands of international importance. A newly appointed RAMSAR site is the Las Piñas-Parañaque Critical Habitat and Ecotourism Area near Manila. Black-winged Stilts and other migratory birds are yearly visitors to the site. Another newly discovered site for migratory waders is the Tibsoc mudflats in Negros Occidental, where large numbers of migrant birds have been seen in recent years. There is also still a big percentage of coastal areas left undiscovered for birdwatching.

FRESHWATER LAKES AND MARSHES

There are numerous freshwater habitats in the Philippines. Some lakes have already been identified as good birdwatching sites, including Naujan Lake in Oriental Mindoro, Malasi Lake in Isabela, and Paoay Lake in Ilocos Norte. There are also marshes and swamps that serve as homes for a big population of birds. The Agusan Marsh in Mindanao is a large wetland almost the size of Metro Manila. It is a core nesting ground for Philippine Ducks, particularly during the wet season when the marsh waters expand, and is also a wintering spot for migratory birds during the dry season. The Candaba Swamp in Luzon has been declared a bird sanctuary after record numbers of migratory birds have been regularly seen there. A number of rare sightings have been recorded in Candaba, including the globally endangered Black-faced Spoonbill.

Candaba Swamp, Luzon

OFFSHORE ISLANDS

Many offshore islands surround the larger mainland areas. The most famous group of these islands is in the Tubbataha Reefs Natural Park in the middle of the Sulu Sea. It has been declared a protected area and UNESCO World Heritage Site. Home to around 1,000 marine species, the islands also serve as an important site for breeding colonies of seven seabird species, including Red-footed Boobies and Black Noddies.

Aside from these habitats, other places that are good for birdwatching include parks and green spaces in urban areas, which have become sites for certain forest-bird species and some migratory birds; grassland in undeveloped areas still serves as habitat for grassland bird species.

Bird Migration and the East Asian Flyway

The Philippines is part of the East Asian-Australasian Flyway. Close to 500 migratory bird species pass through this flyway annually. It encompasses the northern regions of Arctic Russia, including a portion of North America, particularly Alaska, and extends all the way to the southern countries of Australia and New Zealand. There are 37 countries in the flyway, with the Philippines located in the central portion. These countries provide important stopover sites for the millions of migratory birds plying the route. At peak migration months, thousands of waders can be seen in a single location, where the birds rest and feed before continuing with their journeys.

Aside from the many waders and duck species that migrate through the flyway, numerous raptors and passerine species also do so. Thousands of raptors, including Grey-faced Buzzards and Chinese Sparrowhawks, can be seen in strategic parts of the Philippines at peak migration periods. Some passerine birds, such as Arctic Warblers and Grey-streaked Flycatchers, can also be seen in many places, including green spaces in urban areas.

Conserving the Region's Birdlife

The Philippine environment and its wildlife face many threats. One of the biggest is habitat loss. Large amounts of forest cover have been destroyed throughout the years. This deforestation is largely caused by illegal logging and mining. Despite a total log ban declared by the government in 2011, trees are still being cut down and forests are still being destroyed in different parts of the country. Deforestation is also caused by agricultural expansion. The traditional farming practice of slash and burn, locally referred to as *kaingin*, remains a common practice among many farmers. It involves the cutting of trees and burning the area to create patches of farmland where there was none. Traditionally, *kaingin* is not done on a permanent basis. After the harvest the farmers move on to another location to allow the forest and land to regenerate. Unfortunately, this practice has not become the norm. With more and more people needing farmland to cultivate, *kaingin* sites have become permanent areas for most farmers. This has turned the practice into a largely destructive one that is now leading to a large scale of habitat loss.

Another threat to birdlife and habitats is the conversion of wetlands. Large tracts of wetlands and grassland have been converted into farmland, residential communities and commercial areas. Over the years, the numbers of migratory birds have dwindled in areas where the marshes and swamps have been transformed into rice fields or covered up for some other form of development. Reclamation also remains a big threat to coastal habitats. The Las Piñas-Parañaque Critical Habitat and Ecotourism Area in Manila Bay is the last stand of mangroves in the area and is being threatened with reclamation plans. This site is home to large numbers of terns, egrets and other waterbirds. The loss of the habitat would be crucial to the species that migrate there annually.

Hunting is a major threat. It is illegal in the Philippines, but even with legislation in place, hunting and poaching are still prevalent in many areas. Hunting birds for food and for the illegal pet trade are the main reasons behind the continued practice of poaching wild animals from forests and other habitats. Due to unsustainable population growth, many people have resorted to hunting wild birds for food. It has become an easy and cheap livelihood, but can have a devastating effect on already declining bird populations if not controlled. Some people still engage in hunting for sport despite the laws against it, and targets of the hunters are the endemic Philippine Duck, the only endemic duck species in the Philippines, listed as Vulnerable.

The lack of proper law enforcement remains a challenge for environmental conservation in the country. Even with the total log ban, many forests are still being destroyed on a daily basis. Without laws and guidelines in place, people have been logging forests unabatedly, denuding large patches of forest continuously. The effect of the deforestation has devastating effects on bird populations, as does hunting.

Efforts by conservation groups and non-governmental organizations are aimed at helping the implementation and enforcement of existing laws to protect the Philippine environment. Birdwatching is slowly gaining popularity and has, in some cases, become an avenue to re-educate people about being more responsible, and enjoying and making a living out of the environment. It has become a sustainable livelihood for some local guides and communities, who have come to see the business side of being in a birdwatching site frequented by birdwatchers. However, there is a need for support from the local and national government if the threats are to be managed and eradicated. Many of the wildlife species in the Philippines are threatened and will continue to be if nothing is done to protect them and conserve the habitats they live in.

Birdwatching Opportunities in the Philippines

With more than 680 species of bird and high levels of endemic avifauna, the Philippines is an exciting birding destination. Each island has its own specialties, and to see over a hundred of the endemics at least 2–3 weeks must be allowed for intensive birdwatching. A representation of different habitats – coastal, lowland and montane forests, wetlands and grassland – should be visited across the major islands to get a good representation of the diverse bird species the country can offer. Listed below are some of the most popular and accessible birding sites. The adventurous may also like to visit the smaller, more remote islands to see some of the single-island restricted species.

LUZON

Metro Manila This is a highly urbanized and congested area but retains some areas for birdwatching. Quezon City in northern Metro Manila houses sprawling university grounds and small parks, where common birds like Yellow-vented Bulbul, Collared Kingfisher, Golden-bellied Gerygone, Black-naped Oriole, Pied Triller, Philippine Pygmy Woodpecker and Philippine Pied Fantail can be seen. A visit to La Mesa Ecopark in Fairview can give good glimpses of the once-elusive endemic Ashy Thrush, and other species that might be harder to spot elsewhere, including Red-bellied Pitta, Hooded Pitta, Slaty-legged Crake and Indigo-banded Kingfisher.

Manila Bay Coastline Manila Bay, on the western side of Luzon, serves as a central port area for the whole island. Once offering a vast mangrove cover, most of its coastline is now devoted to commerce, and has been developed and reclaimed, or converted into small- to medium-scale fish pens. Several sites along the coast host a significant number and variety of migratory birds such as Common Greenshank, Pacific Golden Plover, Whiskered Tern, Black-headed Gull, Kentish Plover, Black-winged Stilt, Great, Intermediate and Little Egrets, and other species of wader and waterbird. The Las Piñas-Parañaque Critical Habitat and Ecotourism Area (LPPCHEA) in Metro Manila, and Balanga Wetland and Nature Park (BWNP) in Balanga City in Bataan, are two of the sites where infrastructure to support birding activities is available. They can yield rare migrants, pelagic species blown inland by storms and new records. The coastal municipalities of Hagonoy in Bulacan, and Masantol and Sasmuan in Pampanga, where the Pampanga River drains into the bay, are less-explored areas that have significant potential as observation sites during the migration season.

Subic Bay Forests Once an American naval base, the Subic Bay Metropolitan Area houses one of the few remaining lowland forests of Luzon Island. It straddles the provinces of Bataan and Zambales, a three-hour drive from Metro Manila. Many areas have restricted access and permission to visit must be obtained from the SBMA Ecology Centre. Target species here include Luzon Hornbill, Rufous Coucal, Philippine Serpent Eagle, Philippine Hawk-Eagle and Blackish Cuckooshrike. It is also possible to spot the rare White-lored Oriole and White-fronted Tit. A good representation of woodpeckers (Luzon Flameback, Sooty Woodpecker, White-bellied Woodpecker), as well as parrots (Green Racket-tail, Blue-naped Parrot, Guaiabero, Colasisi) and doves (Green Imperial Pigeon, Philippine Green Pigeon, White-eared Brown Dove), can be seen easily here.

Trail in Subic Bay Forests, Bataan

Mt Polis In the Mountain Province of the Cordillera region, this site is fairly accessible for viewing Luzon montane birds. It is located near a popular tourist destination, Banaue, which is well known for its rice terraces. Unfortunately, the forest area, mostly pine and mossy oak, is rapidly shrinking, but birding by the roadside or along vegetable patches can still be productive. Among expected species are Luzon Water Redstart, Mountain Shrike, Flame-breasted Fruit Dove, Luzon Scops Owl, Philippine Bush Warbler and Chestnut-faced Babbler. Mixed flocks can contain Elegant Tit, Warbling White-eye, Blue-headed Fantail, Sulphur-billed Nuthatch and Flame-crowned Flowerpecker.

Mt Makiling Access to Mt Makiling is through the university town of Los Baños in Laguna Province. It is a short drive from Metro Manila and can provide productive birding for a day trip. The lower levels of the mountain can be accessed on foot; a jeepney can be hired to reach the higher elevations of more than 500m. Balicassiao, Scale-feathered and Rough-crested Malkohas, Philippine Bulbul, Yellow-wattled Bulbul, Spotted Wood Kingfisher, White-browed Shama and Grey-backed Tailorbird may be seen here with more ease than at Subic Bay, which largely overlaps Mt Makiling in habitat. Smaller birds such as sunbirds (Flaming, Grey-throated, Handsome, Purple-throated), flowerpeckers (Bicolored, Orange-bellied, Buzzing), and Lowland and Yellowish White-eyes are relatively easy to spot. The surrounding area houses large tracts of agricultural-research rice fields, where Spotted Buttonquail, Barred, Slaty-breasted and Buff-banded Rails, and Plain Bush-hen can be found.

Trail up Mt Makiling, Laguna

Candaba The Candaba Wetlands is a remnant of the once-vast marshlands of central Luzon, which have mostly been converted to rice fields. This is an excellent place to find the endemic Philippine Duck and resident Wandering Whistling Duck. During the peak migratory season in October–February, it is good for large congregations of migratory ducks and waders.

Sierra Madre The Sierra Madre mountain range is the longest in the Philippines, extending from Cagayan province in the north down to the southern province of Quezon. Access to the higher elevation mountains is easiest through the city of Tuguegarao. There is no accommodation and birding here involves setting up camp in the forest. Most target birds here are difficult to spot elsewhere, including the beautiful Whiskered Pitta, Rufous Hornbill, White-lored Oriole, Grand Rhabdornis, Luzon Bleeding Heart, Blue-breasted and Furtive Flycatchers, and Golden-crowned and Luzon Striped Babblers. At fruiting trees a variety of doves may be seen: Cream-bellied Fruit Dove, Amethyst Brown Dove, Black-chinned Fruit Dove and Spotted Imperial Pigeon.

PALAWAN

The avifauna of the Palawan islands group is surprisingly distinctive from that of the rest of the Philippines, reflecting links to the Malaysian faunal regions. Many birds found in Borneo also occur in Palawan but nowhere else in the Philippines, such as Hill Myna, Great Slaty Woodpecker, Blue-eared Kingfisher and Common Iora. Palawan also hosts 20 endemic species that are restricted to the islands, including Palawan Hornbill, Palawan

Peacock Pheasant, White-vented Shama, Palawan Flycatcher, Palawan Scops Owl, Palawan Tit, Spot-throated Flameback, Melodious Babbler, Falcated Wren-Babbler and Blue-headed Racket-tail. Several bulbuls, including Ashy-fronted and Palawan Bulbuls, are also endemic to the island. Palawan is still well forested and birds are abundant. It is also a popular tourist destination so reasonably accessible.

One of the best birding areas is the Puerto Princesa Underground River Subterranean Park (formerly called St Paul's Underground River Subterranean Park), accessed from the village of Sabang. It is situated on the west coast of the island and is iconic of the islands of Palawan: there are large dipterocarp coastal forests, magnificent limestone cliffs and extensive beaches dotted by mangroves. Inland, the Iwahig Penal Colony is a jump-off point to the Balsahan trail, which crosses several inland rivers and lowland forest. The island of Rasa near the town of Narra, south of Puerto Princesa, houses the roosting site of the largest remaining colonies of Red-vented Cockatoo, and is also reliable for large flocks of Pied Imperial Pigeon and the much rarer Grey Imperial Pigeon. More accessible via the tourist route are the islands of Honda Bay, which can offer small island specialists.

MINDANAO

Mt Kitanglad Range Mt Kitanglad in northern Mindanao is the most accessible place for Mindanao montane birds. It is accessed through the province of Bukidnon, and a camp has been conveniently and permanently set up for birders. It is currently the most reliable place to see the Philippine Eagle in the wild, but is also good for several montane species like Apo Myna, Red-eared Parrotfinch, Cinnamon Ibon, Black-and-cinnamon Fantail, Mindanao White-eye, White-cheeked Bullfinch, McGregor's Cuckooshrike, Pinsker's Hawk-Eagle and Philippine Honey Buzzard. Bukidnon Woodcock can often be observed engaging in its roding display by the campsite at dusk. Higher elevations can yield Apo Sunbird, Goodfellow's Jungle Flycatcher and Mountain Shrike. The habitat is a mosaic of forested area, often interrupted by small vegetable fields, but even patches of forest can yield Mindanao Racket-tail, Mindanao Hornbill, Blue-capped Kingfisher and smaller birds like Grey-hooded Sunbird, Rufous-headed Tailorbird, Brown Tit-Babbler, White-browed Shortwing, and Olive-capped and Fire-breasted Flowerpeckers. In the evening nocturnal birds like Philippine Frogmouth, Giant Scops Owl and Everett's Scops Owl may be heard around the lodge and camp.

Mt Talomo This is a couple of hours' drive from downtown Davao City and just a few kilometres outside Mt Apo National Park. It is accessible via a popular resort in Barangay Eden. It is a good alternative for high-elevation birds that are not easily seen in the more popular Mt Kitanglad birding area, including the rare Whiskered Flowerpecker and Cryptic Flycatcher. Other birds include Philippine Needletail, the endemic subspecies of Scarlet Minivet and Pinsker's Hawk-Eagle. Higher elevations might contribute Mindanao Lorikeet and Apo Myna. Occasionally, Philippine Eagle is also seen.

PICOP Bislig in Surigao del Sur Province once housed a large logging concession, but since its closure the lowland forest has fallen prey to illegal logging and settlers. It is still a good area for observing lowland Mindanao forest species, although decent birding areas are getting harder to come by. Logging roads make the area convenient to navigate, although access via hired jeepney can make for a very dusty and bumpy ride. Mindanao, Rufous and Writhed Hornbills still occur in fairly large flocks here. The main targets include striking endemics such as Azure-breasted Pitta, Wattled Broadbill, Black-faced Coucal and Celestial Monarch. More common sightings include species such as Short-crested Monarch, Rufous Paradise Flycatcher, Silvery Kingfisher, Rufous-lored Kingfisher, Philippine Leafbird, Yellowish Bulbul and Philippine Oriole.

One of the trails in PICOP, Bislig

Pasonanca, Zamboanga The Zamboanga Peninsula, which comprises a region of the Greater Mindanao faunal area, hosts its own unique avifauna. Pasonanca Park, half an hour's drive from downtown Zamboanga City, is in the watershed area for the city and contains good forests for Zamboanga specialties such as White-eared Tailorbird and Zamboanga Bulbul. Also seen are various flycatcher species, including Little Slaty Flycatcher and Rufous-tailed Jungle Flycatcher. More remote is Baluno, which also offers a more forested habitat.

MINDORO

Mindoro Island, south-east of Luzon, has never been connected to any of the larger land masses and thus has its own unique set of endemics. The best area for birding lies on the western side of the island, where the Sablayan Penal Colony houses good lowland forest. Mindoro endemics such as Black-hooded Coucal, Scarlet-collared Flowerpecker, Mindoro Bulbul, Mindoro Racket-tail, Mindoro Hornbill and Mindoro Hawk-Owl are concentrated here. The Mindoro Bleeding-Heart, a frequent victim of illegal trapping, might also be seen. On the western side of the island Naujan Lake is not as popular a birding destination, but is known to host migratory ducks at the appropriate season and has been declared a RAMSAR Wetlands Site. Less explored on Mindoro are high-elevation habitats such as Mt Halcon, the highest peak on the island, and Mt Iglit-Baco, which is more popular as the habitat of the highly endangered Philippine Tamaraw – a small water buffalo that is the country's largest endemic mammal. Montane Mindoro birds such as Mindoro Imperial Pigeon and Mindoro Scops Owl can be found at these elevations, but have rarely been seen in recent years.

BOHOL

Bohol is part of the geopolitical region of the Central Visayas; in terms of avifauna it is closer to Mindanao and the eastern islands of Leyte and Samar. The Rajah Sikatuna National Park (RSNP) lies towards the centre of the island and comprises mostly good secondary growth lowland forest. The well-maintained park is located near the popular tourist destination of the Chocolate Hills, a unique geological formation of limestone karst hills that dot the flat landscape. Azure-breasted Pitta or Visayan Broadbill may be seen here. Yellow-breasted Tailorbird is also a target, besides Samar Hornbill, Philippine Trogon, Black-crowned Babbler, Streaked Ground-Babbler, Visayan Blue Fantail, Rufous-lored Kingfisher, Black-chinned Fruit Dove, Philippine Leaf Warbler and Northern Silvery Kingfisher. Everett's Scops Owl and Philippine Frogmouth are good species to be spotted at dusk or dawn. Notable mammalian species that might be seen in the park are *Colugo* or Flying Lemur, and the tiny Philippine Tarsier. The Tarsier Sanctuary in Corella is worth a visit as it also has a small adjacent forested area that is good for birding.

CEBU

Cebu, a long, thin island in the Central Visayas, is the hub of commerce and education in the Visayas region. Cebu City is the highly urbanized capital of Cebu. Much of the island's forest has disappeared, but tiny remnant patches hugging the central mountain ranges still remain and they house two of Cebu's endemic species: Black Shama and the recently rediscovered Cebu Flowerpecker, once thought to be extinct. These, along with Cebu Hawk-Owl, can be seen in the small forests of Tabunan, a couple of hours' drive from Cebu City. Other notable subspecies are the white-bellied form of Balicassiao, the red-headed form of Coppersmith Barbet, Elegant Tit, White-vented Whistler and the rare Streak-breasted Bulbul. The town of Alcoy, around 90km south of Cebu City, is also a good birding area.

The Olango Island Wildlife Sanctuary on the eastern side of Mactan in Metropolitan Cebu is a designated RAMSAR Wetlands site. Its expansive tidal flats and reefs are excellent for shorebirds and egrets during the migration season, and it is one of the most important areas in the country for large congregations of migratory waterbirds. Its flagship species include the globally threatened Chinese Egret and Asian Dowitcher.

Entrance to Balinsasayao Lake, Dumaguete

NEGROS

Twin Lakes and Mantiquil North of the city of Dumaguete are the twin lakes of Balinsasayao and Danao. This area is among the larger tracts of forests on the south-eastern side of Negros Island. Along the perimeter of Lake Balinsasayao the trail can hold sightings of Flame-templed Babbler, Visayan Hornbill, Magnificent Sunbird, Maroon-naped Sunbird, White-winged Cuckooshrike and Yellow-faced Flameback. Mantiquil, south-west of Dumaguete near the town of Siaton, provides access to high-elevation forest that is reliable for Negros Striped Babbler. The Negros subspecies of White-browed Shama may be encountered en route.

Mt Kanlaon Kanlaon Volcano is one of the most active volcanoes in the country and the highest peak on Negros island, rising to almost 2,500m. Its slopes host a large geothermal plant and the resort town of Mambucal. Aside from the birds that might be seen at Dumaguete, the forest here holds Visayan Fantail, Balicassiao, White-browed Shama, White-vented Whistler and Pink-bellied Imperial Pigeon. This is also the home of the elusive Negros Bleeding-heart.

Tibsoc The mudflats at Tibsoc, San Enrique, Negros Occidental host a multitude of migratory waterbirds, including Black-tailed Godwit, Asian Golden Plover, and Great and Red Knots. Less common species like Nordmann's Greenshank have also been spotted here.

The above is but a sampling of the many birding sites in the Philippines. Several other islands and regions are not quite as popular for birding because of logistic limitations, but nonetheless can reveal many birding jewels. The Batanes and Babuyan islands in the north, Camiguin Sur, Siquijor, the Romblon-Tablas-Sibuyan island group and the Sulu archipelago are just some of the islands that have their own specialties. Even on the larger islands, many more remote habitats await the most adventurous birders. They include the Agusan Marshes in Central Mindanao, and the mountains of Mindoro and Palawan; even in Luzon there are still many areas of the Cordilleras and the Sierra Madres that remain to be extensively explored.

Nomenclature and Taxonomy

The checklist follows the taxonomical sequence and nomenclature recommended by the International Ornithologists' Union (IOC) version 10.1. The species selection is based on the *Wild Bird Club of the Philippines (WBCP) Checklist of Birds of the Philippines* (2020). Bird classification is a dynamic process and bird checklists are constantly being revised and updated. With the routine use of molecular techniques and DNA analysis, it is expected that continuous taxonomic rearrangements and revisions of bird names will occur.
The use of bird names in the local dialects can be very confusing; the same bird may have the same or a different name depending on what part of the country you are in. Even more perplexing, the same name may refer to a different bird in another dialect. On top of this, the several minority groups will have their own dialects and their own names for birds. A comprehensive reference on local bird names has yet to be written. It is best to consult residents or local guides, armed with photographs or a field guide, and ask them for information on bird names in an area.

Philippine Megapode ■ *Megapodius cumingi* 35cm

DESCRIPTION Large, ground-dwelling bird with olive-brown upperparts and dark greyish-brown underparts. Small head with bare pinkish eye-skin and dirty-yellow bill. Strong, greyish to dark horn legs. **DISTRIBUTION** All over the Philippines. Most common on isolated islands. **HABITS AND HABITATS** Inhabits coastal scrub, preferring isolated and small islands, and beach forest. Very shy and secretive; often runs away at first sign of disturbance. May also fly a short distance when disturbed. Powerful, strong legs built for foraging on the ground and for piling up loose sand or vegetation where birds lay eggs in communal nests. **SITES** Puerto Princesa Underground River Subterranean Park (Palawan), Apo Island (Mindoro).

Red Junglefowl ■ *Gallus gallus* 66cm

DESCRIPTION Large, ground-dwelling bird very similar to domestic chicken. Male has bare head with distinct red wattle. Neck red-orange streaked with white. Tail feathers long with two central feathers being significantly longer than the rest. Wings black with dark brown edges. Female has variable plumage but is generally dark brown, with yellowish streaks on neck. Skin around eye pink. Female lacks long tail feathers of male. **DISTRIBUTION** Throughout the Philippines. **HABITS AND HABITATS** Relatively hard to find in populated areas where domestic chickens are bred; instead inhabits forests up to 2,000m. Usually heard crowing from vegetation, sounding very similar to domestic chicken, but trails off at last note. **SITES** Subic Forest (Bataan).

Male

Palawan Peacock Pheasant ■ *Polyplectron napoleonis* 40cm ⓔ

DESCRIPTION Large, ground-dwelling bird. Male is stunningly plumaged. Head strikingly patterned black and white, with long iridescent blue-green crest; neck, breast and upper back velvety-black. Lower back, rump and tail black with golden specks, and wings metallic blue-green. Tail adorned with two rows of blue-green 'eyes'. Female mostly brown, with mottled black and off white, but shares white eyebrow and ear-patch of male. **DISTRIBUTION** Palawan. **HABITS AND HABITATS** Very secretive Palawan endemic, preferring lowland forests and secondary growth below 1,000m. Very shy and will walk away at any sign of disturbance. **SITES** Puerto Princesa Underground River Subterranean Park (Palawan). **CONSERVATION** Vulnerable due to limited range, and heavily hunted for illegal wildlife trade.

Male

Wandering Whistling Duck ■ *Dendrocygna arcuata* 49cm; WS 76cm

DESCRIPTION Medium-sized whistling duck. Majority of head and body reddish-brown, contrasting with black crown. Bill black. Wings and back blackish-brown with bold white stripes on flanks. **DISTRIBUTION** Throughout the Philippines. **HABITS AND HABITATS** Common inhabitant of freshwater wetlands and marshes, swamps and rice fields. Often in big flocks that are very noisy when flying, giving characteristic whistling. In flight, tends to hold head down slightly. **SITES** Candaba Marsh (Pampanga).

Philippine Duck ■ *Anas luzonica* 51cm; WS 84cm ⓔ

DESCRIPTION Medium-large dabbling duck. Rusty-coloured head punctuated with dark brown eye-stripe and crown. Greyish-brown body with back and rump darker than breast. Speculum bright green bordered by black. **DISTRIBUTION** Throughout the Philippines. **HABITS AND HABITATS** Favours freshwater lakes, marshes and rivers, but can also be seen in secluded bays and ocean coves. Usually in small groups. In flight, tends to hold head down slightly, in similar way to Wandering Whistling Duck (see p. 17). **SITES** Candaba Marsh (Pampanga), Subic Forest (Bataan), Bislig Airfield (Surigao del Sur).

Yellow Bittern ■ *Ixobrychus sinensis* 33cm; WS 46cm

DESCRIPTION Small bittern. Male mostly buff-brown, darkest on upperparts, with solid dark crown and nape. Female has crown streaked with black and upperparts streaked with brown. Small and squat. Long bill with upper mandible darker brown and lower mandible yellow. Yellow eyes. Characteristic contrast between black outer wing and buff inner wing in flight. **DISTRIBUTION** Throughout the Philippines. **HABITS AND HABITATS** Favours inland freshwater marshes, wetlands and rice fields with reed beds. Solitary and secretive, often not moving and standing on tall grasses and reeds, waiting for prey. **SITES** Candaba Marsh (Pampanga), Bislig Airfield (Surgao del Sur).

Female

Black Bittern

Ixobrychus flavicollis 58.5cm

Male

DESCRIPTION Medium-sized bittern. Male mostly all black on upperparts. Throat white with black mesial stripe. Sides of neck yellowish with dark stripes running down length of neck. Breast black, streaked with white and brown. Female brownish instead of black. Bill long and slender; upper mandible can be dark brown to reddish-brown, and lower mandible varies from green to purple. Yellow to red eyes surrounded by greenish skin that can turn pink. **DISTRIBUTION** Throughout the Philippines, except on Panay, Masbate, Leyte and Bohol. **HABITS AND HABITATS** Mostly crepuscular and solitary. Prefers to stay in reeds and tall grasses in rivers, ponds and marshes. **SITES** Candaba Marsh (Pampanga), Agusan Marsh (Agusan del Sur).

Black-crowned Night Heron

Nycticorax nycticorax 56cm; WS 112cm

DESCRIPTION Large, stocky heron with short neck. Adult birds have black crown with two long white plumes. Upperparts grey, including sides of head, wings and tail. Underparts white. Eyes red and skin around eyes yellow-green. Immature birds have streaked upperparts with white spots, and white underparts with brown streaks. **DISTRIBUTION** Throughout the Philippines. **HABITS AND HABITATS** Crepuscular bird that roosts in trees and other plants near waterbodies during the day. Flies out to feed during the night, waiting at the water's edge and plucking fish out of the shallows. Can be found in rice fields, marshes and coastal areas. **SITES** LPPCHEA (Metro Manila), Candaba Marsh (Pampanga).

Adult

Adult

Nankeen Night Heron

■ *Nyticorax caledonicus* 64cm; WS 116cm

DESCRIPTION Large heron with stocky build and short neck. Adult birds have rufous face, neck and breast, contrasting with black crown and hindneck. Wings darker rufous and feet yellow-green. Immature birds similar to more common immature Black-crowned Night-heron (see p. 19), with upperparts heavily streaked buff and brown. **DISTRIBUTION** Throughout the Philippines. **HABITS AND HABITATS** Mostly crepuscular and may feed in small groups. Roosts in trees in small groups during the day. Prefers freshwater marshes and rice fields, but also recorded near coastal mangroves. **SITES** LPPCHEA (Metro Manila), Olango Island (Cebu).

Adult

Striated Heron

■ *Butorides striata* 42cm; WS 63cm

DESCRIPTION Small heron with mostly dark grey plumage. Black cap and long, dark greenish crest, usually held flat to body. Dark wings have white fringing; underparts buff and brown with white streaks on neck and breast. Black upper bill and yellow lower mandible. Yellow legs. Immature an overall brown with more heavily streaked underparts. **DISTRIBUTION** Throughout the Philippines. **HABITS AND HABITATS** Commonly found in mangroves, mudflats and other coastal areas. Shy and solitary; hunts from rocks in coastal areas or from low-lying branches in mangroves, and rivers and streams. Flies low over water. **SITES** LPPCHEA (Metro Manila), Balanga (Bataan), Olango Island (Cebu).

Javan Pond Heron

■ *Ardeola speciosa* 45cm; WS 81cm

DESCRIPTION Small heron with all-white wings contrasting with buff body in flight. Breeding birds have orange-buff head and neck, transitioning into chestnut at lower neck and breast. Bill yellow with black tip. Non-breeding birds have olive-brown head and neck, with broadly streaked breast; very difficult to separate from non-breeding **Chinese Pond Heron** A. *bacchus*. **DISTRIBUTION** Range has expanded considerably and species can now be found throughout the islands. **HABITS AND HABITATS** Prefers wet rice fields and marshes, and stands motionless while waiting for prey. **SITES** Candaba Marsh (Pampanga), Bislig Airfield (Surgao del Sur), rice fields throughout the Philippines.

Non-breeding

Eastern Cattle Egret

■ *Bubulcus coromandus* 48cm; WS 92cm

DESCRIPTION Medium-sized, stocky white egret with thick, stout head and neck. Breeding birds have orangey head, neck and breast, and long, pinkish-orange plumes on back. Non-breeding birds plain white. Yellow bill and black legs. **DISTRIBUTION** Throughout the Philippines. **HABITS AND HABITATS** Common in rice fields in groups and often follows cattle, water buffalos and horses, hunting the insects disturbed by these large domesticated animals. Sometimes lands on backs of water buffalos. Resident population breeds June–August; augmented September–April by migrant populations from Japan, Taiwan and China. **SITES** Candaba Marsh (Pampanga), Bislig Airfield (Surgao del Sur), rice fields throughout the Philippines.

Breeding

Adult

Grey Heron

Ardea cinerea 101cm; WS 178cm

DESCRIPTION Large, mostly grey heron. Wings and back grey contrasting with paler grey neck and white head and crown; black stripe above eye extends into long black plumes. Black wing edges, with greyish inner wing and contrasting white belly and rest of underparts in flight. Yellow bill and yellow-green legs. **DISTRIBUTION** Throughout the Philippines. **HABITS AND HABITAT** Previously uncommon migrant that has been steadily increasing in numbers in recent years. Favours freshwater marshes as well as tidal flats. Main diet is fish, but will also eat small rodents and frogs. **SITES** Candaba Marsh (Pampanga).

Great-billed Heron

Ardea sumatrana 115cm; WS 190cm

DESCRIPTION Biggest heron in the Philippines. Tall and bulky with mostly dark greyish-brown body and huge black bill. White throat, short crest and grey feet. Uniform grey plumage with no contrast between flight feathers and coverts when in flight. **DISTRIBUTION** Palawan, Negros, Mindoro, Bohol. **HABITS AND HABITAT** Uncommon heron found singly or in pairs in coastal areas, offshore tidal flats and isolated islands. Hunts mainly fish and crustaceans. Stands motionless for long periods while waiting for prey. **SITES** Rasa Island (Palawan), Honda Bay (Palawan).

Adult

Purple Heron

Ardea purpurea 92cm; WS 140cm

Adult

DESCRIPTION Large heron with rufous neck adorned with black stripes, black crown and crest, and dark grey back and wing-coverts. Bill two-toned with black upper mandible and yellow lower mandible. Legs brownish-yellow. Immature has wings and upperparts also rufous toned. **DISTRIBUTION** Throughout the Philippines. **HABITS AND HABITAT** Common heron found singly or in small groups, mostly in freshwater wetlands such as rice fields and marshlands, but also in mangroves. Feeds on fish, frogs and small vertebrates. Flies slowly, with necked retracted in typical heron fashion. Nests in colonies in patches of tall reeds and mangroves. **SITES** Candaba Marsh (Pampanga), rice fields throughout the Philippines.

Intermediate Egret

Ardea intermedia 70cm; WS 102cm

DESCRIPTION Medium-sized white egret. Differs from Eastern Cattle Egret (see p. 21) in larger size and longer neck, and from superficially similar **Great Egret** *Ardea alba* by shorter and thicker, 'S'-shaped neck lacking a kink, and black line of gape ending directly below eye. Yellow bill with black tip; black legs and feet. **DISTRIBUTION** Throughout the Philippines. **HABITS AND HABITATS** Locally common, most often in freshwater marshes, rice fields and inland lakes; sometimes also in mudflats or brackish wetlands. Feeds on fish, frogs, crustaceans and insects. **SITES** Candaba Marsh (Pampanga), other wetlands throughout the Philippines.

Little Egret ■ *Egretta garzetta* 61cm; WS 97cm

DESCRIPTION Small-medium white egret. Breeding birds have two long feathers forming plumes, which are absent in non-breeding birds. Black legs with contrasting yellow feet. Black bill and greenish-yellow facial skin. **DISTRIBUTION** Throughout the Philippines. **HABITS AND HABITATS** Common migrant found in all types of wetland, from freshwater marshes to rice paddies and tidal flats. Feeding strategy involves using feet to stir and disturb the water to flush out prey of fish, insects and crustaceans. **SITES** Candaba Marsh (Pampanga) Balanga (Bataan), other wetlands thoughout the Philippines.

Adult

Pacific Reef Heron ■ *Egretta sacra* 58cm; WS 101cm

DESCRIPTION Small-medium egret occurring in two colour phases: dark and all white. Dark phase uniform slaty-grey with short plumes, and sometimes white chin and stripes in throat. Less common white phase all white with plumes. Bill pale yellow, and long and drooping in shape; legs yellowish-green. **DISTRIBUTION** Coastal areas in the Philippines. **HABITS AND HABITATS** Uncommon egret favouring rocky shorelines, exposed reefs and isolated islets. Hunts from the water's edge looking for fish and crustaceans. When flushed flies low over the water. **SITES** Puerto Princesa Underground River Subterranean Park (Palawan), coastal areas thoughout the Philippines.

LEFT: *White phase*; RIGHT: *Dark phase*

Chinese Egret ■ *Egretta eulophotes* 69cm; WS 99cm

DESCRIPTION Small-medium white egret. Breeding-plumage birds have long tufts of white feathers on nape, forming plumes extending up to tail. Yellow bill, bright blue facial skin and all-black legs with yellow feet. Non-breeding birds have black bills, with base of lower bill yellow; legs yellowish-green. Differs from similar white-phase Pacific Reef Heron (see opposite) in more dagger-shaped, rather than drooping bill, longer and more elegantly 'S'-shaped neck, and longer legs. **DISTRIBUTION** Luzon, Palawan, Cebu, Panay, Negros, Bohol, Samar. **HABITS AND HABITATS** Rare non-breeding migrant to the Philippines. Characteristic active feeding strategy whereby it is frequently seen running manically after prey of fish and crustaceans. Mainly found in shallow tidal flats, and retires to mangroves for roosting. **SITES** Olango (Cebu). **CONSERVATION** Classified as Vulnerable migrant with current estimates of 2,500 individuals worldwide. Decline primarily caused by reclamation of its tidal mudflat feeding habitat. Wintering sites on Palawan and in the Central Philippines, particularly Olango Island in Cebu, may form core of its winter range.

Western Osprey ■ *Pandion haliaetus* 56 cm; WS 135cm

DESCRIPTION Large raptor with white-and-brown plumage. White head with brown streaks and short crest. Distinctive black stripe through eye. Upperparts brownish-black. Neck and breast mostly white. Underwings with white coverts contrasting with brownish barred flight feathers. In flight, wings held in 'bent' shape. **DISTRIBUTION** Throughout the Philippines. **HABITS AND HABITATS** Feeds predominantly on fish, and found around inland lakes, large fish ponds and coastal areas. Adept hunter, hovering first, then diving feet first into the water to catch fish in its talons; prey is then brought to a favourite perch, usually a tall dead tree, for consumption. **SITES** LPPCHEA (Metro Manila), Puerto Princesa Underground River Subterranean Park (Palawan), Twin Lakes (Negros).

Black-winged Kite

■ *Elanus caeruleus* 37cm; WS 101cm

DESCRIPTION Small-medium raptor with long wings and striking pale grey-and-black plumage. White head, nape, back and wings grey with distinctive black 'shoulders'. In flight appears almost all white, with edges of primaries tinged with darker grey. **DISTRIBUTION** Throughout the Philippines except Palawan. **HABITS AND HABITATS** Uncommon raptor in grassland and open scrub, where it likes to perch conspicuously on exposed trees or poles. Often hovers over fields looking for prey. Flight is graceful. **SITES** Malasi Lakes (Cagayan), Mt Kitanglad (Bukidnon).

Crested Honey Buzzard ■ *Pernis ptilorhyncus* 59cm; WS 142cm

DESCRIPTION Medium-large raptor characterized by long, pigeon-like neck and long tail with three dark bands. Plumage mostly light brown with different shades of upperparts and underparts for the different races. In flight shows barred flight feathers; bases of primaries translucent, forming 'window'; underparts cinnamon. Eye light yellow. **DISTRIBUTION** Throughout the Philippines. **HABITS AND HABITATS** Mainly feeds on bees gathered from nests in tree hollows, or hives suspended from huge tree branches. Conspicuous when soaring over forest edges in lowlands and middle elevations. **SITES** Tanay (Rizal), Mt Kitanglad (Bukidnon), PICOP (Surigao del Sur).

Crested Serpent Eagle

Spilornis cheela 51cm; WS 107cm

DESCRIPTION Medium-sized raptor with broad white crest, and generally all-brown plumage with white spots on chest. In flight shows diagnostic broad white subterminal band in underwing and tail. Yellow cere and yellow unfeathered feet. Immature mostly white with black border to whole of underwing. **DISTRIBUTION** Palawan. **HABITS AND HABITATS** Common raptor inhabiting forests and forest edges, soaring over and looking for preferred prey of snakes and other small reptiles. When flying wings noticeably bent forwards with slight dihedral. Occurs from lowlands up to montane forests. **SITES** Iwahig (Palawan), Sabang (Palawan).

Philippine Serpent Eagle

Spilornis holospilus 51cm; WS 107cm e

DESCRIPTION Very similar to Crested Serpent Eagle (see above), which it replaces away from Palawan. **DISTRIBUTION** Throughout the Philippines except Palawan. **HABITS AND HABITATS** Most common of all Philippine forest raptors. Mostly seen soaring above forests and forest edges, giving out its distinctive whistled call. Hunts along forest edges for snakes, skinks and other reptiles. **SITES** Subic Forest (Bataan), Mt Makiling (Laguna), PICOP (Surigao del Sur), Mt Kitanglad (Bukidnon).

LEFT & RIGHT: *Adult*

Philippine Eagle ■ *Pithecophaga jefferyi* 99cm; WS 188cm ⓔ

DESCRIPTION Massive raptor with relatively short but broad wings, long tail, and all-white underparts and underwings in flight. At rest white underparts and head contrast with dark brown upperparts and tail. Shaggy crest on head has distinct dark streaking. Huge grey-blue bill and yellow talons. **DISTRIBUTION** Main strongholds are Mindanao and Luzon, with a small number of pairs on Samar and Leyte. **HABITS AND HABITATS** Inhabits forests and forest edges from lowlands to over 2,000m, but now mostly confined to mountains. Hunts from perches within canopy, where rarely seen, but regularly soars above canopy in good weather. **SITES** Mt Kitanglad (Bukidnon), Mt Talomo (Davao), Mt Apo (South Cotabato). **CONSERVATION** National Bird of the Philippines, but sadly Critically Endangered. Population estimated at 50–250 pairs, with continuing declines due to forest destruction and fragmentation, and hunting. Attempts to reintroduce captive-bred individuals have so far had only limited success.

Changeable Hawk-Eagle

■ *Nisaetus cirrhatus* 64cm; WS 127cm

DESCRIPTION Medium-large raptor with very short white crest and feathered tarsus. Occurs in three phases: dark, light and intermediate. Dark phase entirely dark brown. Light phase has dark head with fine streaks on face, and dark brown nape all the way to back and wings. Neck and breast white with dark streaks. Intermediate phase variable in plumage. Immature birds very similar to immature Crested Honey Buzzards (see p. 26) and serpent eagles. **DISTRIBUTION** Palawan, Mindanao, Mindoro. **HABITS AND HABITATS** Uncommon raptor preferring secondary growth and logged forests in lowlands. Perches on exposed branches at forest edges. **SITES** Sabang (Palawan).

Dark phase

Philippine Hawk-Eagle

■ *Nisaetus philippensis* 61cm; WS 119cm ⓔ

DESCRIPTION Medium-large raptor with long crest. Adult birds very dark with mostly dark brown plumage. Crest made up of several long feathers. Dark brown head, neck, back and wings. Neck and small part of chest white with black streaks. Has mesial stripe. Belly reddish-brown with no barring. Bright yellow feet with feathered tarsus. **DISTRIBUTION** Luzon, Mindoro. **HABITS AND HABITATS** Uncommon raptor found in primary and secondary forests from lowlands to montane areas. Often soars when thermals develop and draws attention with distinctive two-note call. **SITES** Subic Forest (Bataan), Mt Makiling (Laguna). **CONSERVATION** Endemic classified as Endangered due to lowland forest loss, hunting and illegal trade contributing to its rapid decline.

Adult

Pinsker's Hawk-Eagle

■ *Nisaetus pinskeri* 60cm; WS 118cm ⓔ

DESCRIPTION Medium-large raptor with long crest. Formerly lumped with Philippine Hawk-Eagle (see above) and very similar in plumage, but does not overlap in range. Differs from Philippine Hawk-Eagle by having lighter underparts, barred brown and black and white belly; slightly smaller in size. **DISTRIBUTION** Mindanao, Negros, Samar, Leyte, Bohol. **HABITS AND HABITATS** Uncommon raptor often found soaring over forests and forest edges. Hunts from concealed perches and prefers to eat small rodents and lizards. **SITES** Mt Kitanglad (Bukidnon), PICOP (Surigao de Sur). **CONSERVATION** Like Philippine Hawk-Eagle, this endemic is classified as Endangered due to lowland forest loss, hunting and illegal trade contributing to its rapid decline.

Adult

Adult, Palawan

Crested Goshawk

■ *Accipiter trivirgatus* 33cm; WS 56cm

DESCRIPTION Medium-small accipiter with short crest. Adults have very dark head, yellow eyes, yellow cere, and white neck with black mesial stripe. Back and wings dark brown. Chest and belly barred brown and white, with different types of barring according to race. Palawan race only lightly barred compared with *extimus* race found in Mindanao. **DISTRIBUTION** Mindanao, Samar, Leyte, Negros, Bohol, Palawan. **HABITS AND HABITATS** Common accipiter preferring lowland forests and forest edges. Hunts from very low in forests and prefers to eat birds, lizards and small rodents. **SITES** Sabang (Palawan), Twin Lakes (Negros), PICOP (Surigao del Sur).

Male, Luzon

Besra

■ *Accipiter virgatus* 26cm; WS 46cm

DESCRIPTION Small accipiter. Dark blue-grey head, yellow eyes and cere, and white throat with characteristic broad dark mesial stripe. Upperparts dark grey; underparts mainly white with fine reddish barring. Vent white. **DISTRIBUTION** Throughout the Philippines except Palawan. **HABITS AND HABITATS** Uncommon raptor of lowlands and montane areas. Prefers to hunt within forest using surprise tactics, flying from concealed perches, then catching prey unawares. Prey includes small birds, reptiles and insects. **SITES** Subic Forest (Bataan), Rajah Sikatuna National Park (Bohol), Mt Kitanglad (Bukidnon).

Eastern Marsh Harrier

■ *Circus spilonotus* 54cm; WS 122cm

DESCRIPTION Medium-sized raptor with long wings and tail. Sexually dimorphic, with male birds having black, grey and white plumage, while female birds are dark brown and white. Male birds have streaked black-and-white heads, necks and throats; black face, back and primaries. Females generally all brown with streaks of white. **DISTRIBUTION** Throughout the Philippines. **HABITS AND HABITATS** Uncommon migrant favouring marshes, rice fields and open grassland. Hunts low over the ground with wings held in shallow 'V' shape, looking for small mammals, lizards, frogs and small birds. **SITES** Candaba Marsh (Pampanga).

Female

Brahminy Kite ■ *Haliastur indus* 51cm; WS 122cm

DESCRIPTION One of the most common medium-sized raptors in the Philippines. Head, upper back, breast and upper belly white; rest of body rufous. Immature birds brown rather than rufous, with pale-buff rather than white head. In flight easily identified, with white flight feathers contrasting with rufous underwing and black wing-tips. **DISTRIBUTION** Throughout the Philippines. **HABITS AND HABITATS** Most common in forested areas near coasts, but can also be found in mountains. Feeds mainly on fish, but will also eat other animals such as snakes, lizards and frogs. **SITES** Subic Forest (Bataan), Mts Palay Palay Mataas Na Gulod National Park (Cavite).

Adult

Philippine Falconet

■ *Microhierax erythrogenys* 16cm; WS 25cm ⓔ

DESCRIPTION Smallest raptor in the Philippines. Characterized by black-and-white plumage, with crown, upperparts, thighs, wings and flanks glossy blue-black, and white face, breast and belly. Brown eyes with black cere, serrated black bill and grey legs. **DISTRIBUTION** Throughout the Philippines except Palawan. **HABITS AND HABITATS** Common raptor preferring open forests and forest edges in lowland and mid-montane areas. Perches on tall snags and dead trees, looking for prey of insects like click beetles and cicadas, lizards and sometimes small birds. **SITES** Subic Forest (Bataan), Mt Makiling (Laguna), PICOP (Surigao del Sur).

Female

Common Kestrel

■ *Falco tinnunculus* 36cm; WS 74cm

DESCRIPTION Small-medium raptor with mottled brown and buff-white plumage. Sexually dimorphic. Male has mottled dark grey crown and nape, yellow eyes and cere, and buff-white throat with black moustachial stripe. Back and upperparts light rufous-brown with small black spots. Light grey, long and pointed wings, and longish tail with black subterminal band. **DISTRIBUTION** Luzon, Mindanao, Palawan. **HABITS AND HABITATS** Uncommon migrant recorded in October–March. Often found in open country, sometimes even in cities and airports. Frequently hovers while looking for prey of small rodents; sometimes catches small bats on the wing. **SITES** UP Los Baños (Laguna).

Peregrine Falcon

■ *Falco peregrinus* 41cm; WS 92cm

DESCRIPTION Large and powerful falcon with dark grey and buff-white plumage. Migrant and resident races occur in the Philippines, with resident races having an almost solid black head, dark brown eyes surrounded by yellow skin, yellow cere and dark grey, black-tipped beak. Back and upperparts sooty-black; throat and breast buff-white with very fine black striations. Powerful yellow legs. Migrant birds look very similar, but have white ear-patch, and lighter underparts with less barring. **DISTRIBUTION** Throughout the Philippines. **HABITS AND HABITATS** Uncommon raptor occurring in different habitats, with resident birds living in forests and forest edges, while migrant birds occur in more open areas, wetlands and coasts, and even on skyscrapers in cities. Very fast and powerful flier, catching prey in mid-air. **SITES** Candaba Marsh (Pampanga), Mt Makiling (Laguna), Mt Kitanglad (Bukidnon).

Adult, Luzon (migrant)

Slaty-legged Crake ■ *Rallina eurizonoides* 27cm

DESCRIPTION Small-medium rail with dark rufous head and neck, orange eyes and bright yellow eye-ring. Upper bill black and lower bill lime green. Back deep olive, neck and upper breast reddish-brown, with rest of underparts black finely barred with white stripes. Grey legs that separate it from similar-looking **Red-legged Crake** *R. fasciata*. **DISTRIBUTION** Throughout the Philippines except Palawan. **HABITS AND HABITATS** Uncommon and very secretive crake preferring dark forest floor and secondary growth. Forages for prey using bill to look for insects and grubs in leaf litter. Partly nocturnal and will run away at the first sign of danger. **SITES** La Mesa Ecopark (Metro Manila), PICOP (Surigao del Sur).

Barred Rail

■ *Gallirallus torquatus* 32cm

DESCRIPTION Medium-large rail with striking black, brown and white plumage. Crown brown with black face accented by white stripe below red eye. Neck and back deep brown, throat black, and rest of underparts black with fine white barring. Also has less distinct brown breast-band. Bill and legs dark horn. **DISTRIBUTION** Throughout the Philippines. **HABITS AND HABITATS** Common but shy rail found in rice fields, wetlands and forest edges. Can be seen in the early morning foraging for insects and grubs in rice paddies, roadsides, flooded wetlands and secondary scrub. **SITES** Candaba Marsh (Pampanga), La Mesa Ecopark (Metro Manila), rice fields and wetlands throughout the Philippines.

Buff-banded Rail ■ *Gallirallus philippensis* 27cm

DESCRIPTION Medium-sized rail smaller than Barred Rail (see above) and predominantly brown, black and white. Reddish-brown crown, white stripe above red eye, rufous-brown band across eye and buff-white throat. Neck white finely barred with black stripes; rest of upperparts mottled brown and black. Also has buff-brown band across chest. Bill grey with pinkish base; legs light brown. **DISTRIBUTION** Throughout the Philippines except Palawan. **HABITS AND HABITATS** Similar to Barred Rail but lighter in colour, and prefers drier areas of rice fields, marshes and grassland. Very shy and hard to flush, and will run away at first sign of danger. Feeds on a variety of food, including insects, small vertebrates, fallen seed and fruits. **SITES** UP Los Baños (Laguna), Candaba Marsh (Pampanga), rice fields and wetlands throughout the Philippines.

Plain Bush-hen ■ *Amaurornis olivacea* 30cm ⓔ

DESCRIPTION Very dark, medium-sized rail with dark grey head, breast and belly, and dark olive-brown upperparts, wings and flanks. Lime green bill, blood-red eyes and olive-yellow legs. **DISTRIBUTION** Throughout the Philippines except Palawan. **HABITS AND HABITATS** Common but very secretive endemic preferring to stay in grassland, scrubs and secondary forest edges. Very distinctive, raspy, growling call is often heard. but most of the time birds stay hidden in dense bush. **SITES** Candaba Marsh (Pampanga), UP Los Baños (Laguna), Mt Kitanglad (Bukidnon).

White-breasted Waterhen ■ *Amaurornis phoenicurus* 28cm

DESCRIPTION Medium-sized rail with white plumage from forecrown up to belly. Rest of upperparts slaty-black, with rump and lower underparts rufous-brown. Dark red eyes, olive-yellow bill with spot of red on upper part and olive-yellow legs. **DISTRIBUTION** Throughout the Philippines. **HABITS AND HABITATS** Common waterbird living in flooded grassland and wetlands; also forest edges and mangroves. The most conspicuous among the rails in the Philippines, and often comes out on roads to forage for food. **SITES** Candaba Marsh (Pampanga), rice fields and wetlands throughout the Philippines.

White-browed Crake

Porzana cinerea 21cm

DESCRIPTION Small rail with dark grey crown, white brow above and below eye, and black stripe through eye. Short, greenish-yellow bill, bright red eyes, and brown upperparts contrasting with lighter greyish underparts. Dull yellow legs. **DISTRIBUTION** Throughout the Philippines. **HABITS AND HABITATS** Common slim-bodied crake inhabiting wetlands, grassland, rice fields, tidal marshes and lakes. Not as shy as most crakes in the Philippines, and can be seen walking on floating vegetation with its long, nimble feet. **SITES** Candaba Marsh (Pampanga), rice fields and wetlands throughout the Philippines.

Philippine Swamphen

Porphyrio pulverulentus 41cm e

DESCRIPTION Large, heavily built waterbird with large purplish-blue head and face accented by triangular bright red bill and frontal shield. Most of body bluish-purple, with back, rump and flight feathers dark olive-green and undertail-coverts white. Big and strong red feet. **DISTRIBUTION** Luzon, Mindoro, Panay, Bohol, Mindanao. **HABITS AND HABITATS** Uncommon rail inhabiting freshwater wetlands and swamps, abandoned rice fields and lakes. Walks on top of vegetation with its big red feet, and prefers a diet of snails and other invertebrates. **SITES** Candaba Marsh (Pampanga), wetlands throughout the Philippines.

Barred Buttonquail

■ *Turnix suscitator* 12cm

DESCRIPTION Medium buttonquail with male and female differing in plumage. Female has black head with tiny white spots above eye and sides of face. Back is rufous with black and brown mottling on upperparts. Buff underparts have thick dark barring. Male lacks black head and rufous back of female. Bill is yellow in females and black in males. Legs are yellow. Eyes yellow. **DISTRIBUTION** Luzon, Palawan, Mindoro, Masbate, Negros, Cebu. **HABITS AND HABITATS** Usually seen in grasslands and even on dirt roads foraging alone or in pairs. When flushed, flies to hide in grass. **SITES** UP Los Baños (Laguna), fields throughout its range.

Female

Black-winged Stilt

■ *Himantopus himantopus* 36cm

DESCRIPTION Large wader with very long neck, long, thin bill, very long pink legs, and black-and-white plumage. Head, nape and underparts all white; wings all black. Non-breeding and immature birds have greyish necks. **DISTRIBUTION** Throughout the Philippines. **HABITS AND HABITATS** Common migrant wader to the Philippines, congregating in both freshwater wetlands like rice paddies, and coastal marshes and abandoned fish ponds. Can be found singly or sometimes in big flocks, foraging for invertebrates and fish. Very graceful in flight, with bright pink legs quite obvious. **SITES** Candaba Marsh (Pampanga), Balanga (Bataan), wetlands and ponds throughout the Philippines.

Pacific Golden Plover

■ *Pluvialis fulva* 23cm

DESCRIPTION Medium-large wader with golden speckles on upperparts. Birds in breeding plumage have bright golden spots on upperparts with clear white line starting from forehead and above eye running down sides just below wing. Face, throat, breast and belly up to vent brownish-black. Non-breeding plumage birds dappled light brown on underparts. Black bill and legs. **DISTRIBUTION** Throughout the Philippines. **HABITS AND HABITATS** Common plover recorded all year round except in July. Occurs in small groups, favouring a wide variety of habitats such as exposed mudflats and coral reefs, rice fields and short grassland. **SITES** Candaba Marsh (Pampanga), Balanga (Bataan).

Grey Plover ■ *Pluvialis squatarola* 28cm

DESCRIPTION Large plover similar to Pacific Golden Plover (see above), but lacks golden spots on upperparts and is bigger. Birds in breeding plumage light brown above with dark spots; white stripe starting from forehead running above eye up to lower breast. Underparts all black. White rump, white wing-bar and black armpits separate this species from smaller Pacific Golden Plover. **DISTRIBUTION** Throughout the Philippines. **HABITS AND HABITATS** Common migrant that prefers to forage on exposed mudflats, isolated beaches and exposed coral flats. Recorded all year except in July. **SITES** Olango (Cebu), Tibsoc Mudflats (Negros), coastal mudflats throughout the Philippines.

LEFT: *Pacific Golden Plover*; RIGHT: *Grey Plover*

Little Ringed Plover

■ *Charadrius dubius* 17cm

Non-breeding

DESCRIPTION Small plover with brown cap, white forehead, white brow above and black band across diagnostic yellow eye-ring. Back and wings brownish-grey, and belly all white with complete black breast-band. Flesh-coloured legs. Immature birds have less distinct yellow eye-ring than adults. **DISTRIBUTION** Throughout the Philippines. **HABITS AND HABITATS** Common wader with both resident and migrant populations. Occurs in small groups, usually in freshwater wetlands, flooded rice paddies and inland drained fish ponds; less so in coastal mudflats. **SITES** Candaba Marsh (Pampanga), Balanga (Bataan), wetlands throughout the Philippines.

Malaysian Plover

■ *Charadrius peronii* 16cm

Male

DESCRIPTION Smallest plover in the Philippines, with similar plumage to **Kentish Plover**'s *C. alexandrinus*. Can be separated from Kentish by pale-tinged upperparts, appearing scaly, thick bill, and complete or nearly complete narrow black breast-band continuing around neck just below white collar. Female has rufous breast-band. **DISTRIBUTION** Throughout the Philippines. **HABITS AND HABITATS** Uncommon plover found on isolated sand and coralline beaches, usually in pairs. One of the few resident shorebirds of the Philippines. **SITES** Sabang (Palawan).

Bukidnon Woodcock

■ *Scolopax bukidnonensis* 33cm ⓔ

DESCRIPTION Large, very plump wader with predominantly reddish-brown plumage. Head a lighter reddish-brown with three darker brown stripes on crown and neck, and black stripe running across eye. Very long, greyish-brown bill. Rest of upperparts darker brown with fine barring; underparts lighter and more buff. **DISTRIBUTION** Luzon and Mindanao. **HABITS AND HABITATS** The only endemic wader in the Philippines. Nocturnal, preferring to live in middle and montane forests above 1,000m. Mostly rests concealed on the ground during the day. At dusk and just before dawn, performs roding display, flying in circuit while calling. **SITES** Mt Kitanglad (Bukidnon), Mt Polis (Mountain Province).

Asian Dowitcher ■ *Limnodromus semipalmatus* 34cm

DESCRIPTION Medium-large wader characterized by very long, straight bill. In breeding plumage predominantly reddish-brown with head, neck, breast and belly dark rufous.Top of head, back and upperparts black with rufous edges. Tail has black and white striations. In non-breeding plumage all grey with upperparts and wings darker in colour. Bill black and straight, and thicker at base and at tip. Dark grey legs. **DISTRIBUTION** Cebu, Luzon, Negros. **HABITS AND HABITATS** Rare and highly localized migrant occurring in coastal mudflats. Characteristic sewing machine-like feeding strategy; forages by probing with long bill in shallows and mudflats, with head and body bobbing up and down like needle on sewing machine. When walking holds long bill downwards. Often seen with similar-looking Bar-tailed Godwits (see opposite). **SITES** Olango (Cebu), Tibsoc Mudflats (Negros). **CONSERVATION** Classified as Near Threatened due to ongoing destruction of suitable wintering grounds.

Non-breeding

Black-tailed Godwit

Limosa limosa 38cm

DESCRIPTION Large wader with very long, straight, two-toned bill. In breeding plumage has reddish-brown head with white stripe above eye, reddish-brown neck and nape, and brown upperparts with black speckles. Underparts white with black striations. In non-breeding plumage generally white and grey, with light brown and grey-speckled upperparts and wings. Best diagnostic features are almost straight, flesh-pink bill with dark brown tip and, in flight, white wing-bars contrasting with grey flight feathers and solid black-tipped white tail. **DISTRIBUTION** Luzon, Negros, Mindoro, Samar, Mindanao. **HABITS AND HABITATS** Rare migrant to the Philippines, preferring to stay in coastal mudflats usually in small groups together with other waders. Has been recorded all year as some birds oversummer. **SITES** Olango (Cebu), Bangrin (Pangasinan), Tibsoc Mudflats (Negros).

Non-breeding

Bar-tailed Godwit

Limosa lapponica 38cm

DESCRIPTION Large wader with very long, slightly upturned bill, fairly long neck and long legs. In breeding plumage generally rufous all over with upperparts darker than underparts. Non-breeding birds grey all over. Easily differentiated from Black-tailed Godwit (see above) by slightly upturned, two-tone bill that is fleshy-pink with black tip. **DISTRIBUTION** Luzon, Samar, Negros, Bohol, Mindanao. **HABITS AND HABITATS** Uncommon migrant preferring coastal area and river deltas in mudflats and exposed coral flats. Often seen feeding with the similar but rarer Asian Dowitcher (see opposite), which has straight black bill. Uses long bill to extract worms from soft mud. Recorded in the Philippines in August–May. **SITES** Olango (Cebu), Tibsoc Mudflats (Negros).

Non-breeding

Eurasian Whimbrel ■ *Numenius phaeopus* 44cm

DESCRIPTION Large and plump wader with dark crown, white eyebrow and black eye-stripe. Medium-length curved bill for probing into mud. Upperparts mottled brown; underparts buff-white with lighter brown barring. **DISTRIBUTION** Throughout the Philippines. **HABITS AND HABITATS** Common and very distinct migrant recorded in the Philippines throughout the year except May. Found in coastal areas feeding on mudflats and sandy beaches in small groups. **SITES** LPPCHEA (Metro Manila), Balanga (Bataan), Olango (Cebu).

Common Redshank ■ *Tringa totanus* 28cm

DESCRIPTION Medium-sized wader with generally white and greyish-brown plumage. Crown greyish with short white eyebrow. Bill two toned with reddish-pink at base and dark brown at tip. Neck and sides of breast finely streaked; upperparts finely barred brown. Distinct white band near edge of black-tipped wing. Orange-red legs. **DISTRIBUTION** Throughout the Philippines. **HABITS AND HABITATS** Common migrant to the Philippines recorded August–June, with some birds oversummering. Prefers coastal areas, feeding in shallows and mud, and sometimes also in wet rice fields. **SITES** LPPCHEA (Metro Manila), Balanga (Bataan), Olango (Cebu).

Non-breeding

Marsh Sandpiper

■ *Tringa stagnatilis* 25cm

DESCRIPTION Medium-sized wader with greyish-brown and white plumage. Small white head has brown streaks on crown and diagnostic straight, very fine, needle-like bill. Upperparts greyish-brown with dark shoulder; underparts all white. Very thin green legs. **DISTRIBUTION** Throughout the Philippines. **HABITS AND HABITATS** Uncommon migrant occurring mostly in freshwater habitats such as flooded rice fields and abandoned fish ponds, but sometimes also in coastal mudflats. Fine thin bill and legs separate it from other similar looking waders. Recorded August–May. **SITES** Candaba Marsh (Pampanga), Balanga (Bataan), wetlands throughout the Philippines.

Breeding

Wood Sandpiper

■ *Tringa glareola* 21cm

DESCRIPTION Medium-sized, slim wader. Dark crown accented by white stripe above eye, extending up to nape. Neck finely streaked with brown stripes. Upperparts dark brown finely mottled with white spots. Underparts and belly all white. Two-toned, thickish bill greenish at base and black at tip. Yellowish legs. **DISTRIBUTION** Throughout the Philippines. **HABITS AND HABITATS** Common wader preferring freshwater marshes such as flooded rice fields, fish ponds and inland wetlands; rarely on coast. Can be found in small groups, and frequently bounces tail and rump up and down. Recorded July–May. **SITES** Candaba Marsh (Pampanga), Balanga (Bataan), rice fields and wetlands throughout the Philippines.

Grey-tailed Tattler ■ *Tringa brevipes* 25cm

DESCRIPTION Medium-sized wader with stocky build. Grey crown and distinctive white eyebrow above thin black eye-stripe. Neck and upperparts uniform grey; belly and underparts all white. Bill two toned, with base of lower mandible yellow and rest of bill black. Legs yellow, and very short and thick compared with those of other waders. **DISTRIBUTION** Throughout the Philippines. **HABITS AND HABITATS** Common wader found in a variety of habitats, including coastal mudflats, sandy beaches and river deltas. Recorded August–May and some birds oversummer. Bobs tail up and down like Common Sandpiper (see below). **SITES** LPPCHEA (Metro Manila), Balanga (Bataan), Olango (Cebu), coastal areas throughout the Philippines.

Non-breeding

Common Sandpiper ■ *Actitis hypoleucos* 20cm

DESCRIPTION Small-medium wader that is quite distinct, having short legs and regular bobbing movement. Generally brown-and-white plumage, with greyish-brown head and face accented by white eyebrow. Neck greyish-brown and upperparts up to wing uniformly greyish-brown. Belly and underparts all white. **DISTRIBUTION** Throughout the Philippines. **HABITS AND HABITATS** Widespread wader occurring in coastal wetlands and mudflats, and freshwater marshes like rice paddies, even on shores of mountain streams and rivers. Easily identified by continuous bobbing of tail while walking, and distinctive stuttering of wings when in flight. Recorded all year round. **SITES** LPPCHEA (Metro Manila), Balanga (Bataan), Olango (Cebu), wetlands, rivers and coastal areas throughout the Philippines.

Ruddy Turnstone

■ *Arenaria interpres* 21cm

DESCRIPTION Small-medium, stocky wader with highly variable black, white and dark reddish plumage. Breeding birds have dark crown, and white face accented by black eye-stripe joining black throat and black breast-band. Upperparts black and reddish-brown; belly and underparts white. Non-breeding birds have mottled brown upperparts and white underparts. Bright orange legs. **DISTRIBUTION** Throughout the Philippines. **HABITS AND HABITATS** Uncommon wader found in coastal areas like exposed mudflats, sandy beaches and coral flats. As its common name implies, turns stones and pebbles with chisel-like bill when foraging. Recorded August–May. **SITES** LPPCHEA (Metro Manila), Olango (Cebu), coastal areas throughout the Philippines.

Great Knot ■ *Calidris tenuirostris* 27cm

DESCRIPTION Medium-large wader with primarily black, grey and white plumage. Very slightly downcurved bill, and finely streaked head and neck. Upperparts and wings have dark mottling edged with white, which appears as spots. Breast and upper belly have lighter spots; lower belly all white. Buff-grey legs. The biggest among the *Calidris* waders. **DISTRIBUTION** Luzon, Negros, Cebu, Palawan. **HABITS AND HABITATS** Uncommon and can be found in coastal areas with exposed mudflats in small groups. Mainly feeds on molluscs and insects. Recorded August–May. **SITES** Olango (Cebu), Tibsoc Mudflats (Negros).

Non-breeding

Red Knot ■ *Calidris canutus* 24cm

DESCRIPTION Medium-sized, stocky *Calidris* wader similar to less common Great Knot (see p. 45). In breeding plumage has reddish-brown face with darker crown, reddish neck and underparts up to belly, and mottled black and reddish-brown upperparts. In non-breeding plumage is grey and white, with grey face and white eyebrow. Upperparts mottled greyish, breast finely streaked and underparts all white. Short, thick bill and grey-buff legs. Distinguished from Great Knot by smaller size, shorter bill and lighter primary coverts. **DISTRIBUTION** Luzon, Mindoro, Palawan. **HABITS AND HABITATS** Uncommon migrant to the Philippines found in coastal areas with sandy beaches and exposed mudflats. Gregarious, and can form large groups in suitable habitat. **SITES** Olango (Cebu), Tibsoc Mudflats (Negros).

Non-breeding

Red-necked Stint ■ *Calidris ruficollis* 16cm

DESCRIPTION Small *Calidris* wader that is one of the smallest migrant waders occurring in the Philippines. Breeding birds generally rufous and dark brown and white. Head and face rufous with an eyebrow, dark crown and short black bill. Upperparts darker brown; underparts all white. Non-breeding birds all grey and white with noticeable white wing-bar. All-black legs. **DISTRIBUTION** Reported throughout most of the Philippines. **HABITS AND HABITATS** Common migrant preferring coastal mudflats and newly ploughed rice fields. Feeds on small crustaceans, insects and snails. Recorded all year round. **SITES** LPPCHEA (Metro Manila), Candaba Marsh (Pampanga), Balanga (Bataan).

Non-breeding

Long-toed Stint

■ *Calidris subminuta* 14cm

DESCRIPTION Small wader very similar to Red-necked Stint (see opposite). Differs from Red-necked by smaller size, overall browner plumage, yellow instead of black legs, and presence of light grey breast-band. Tends to feed in slower, more deliberate manner. **DISTRIBUTION** Reported throughout most of the Philippines. **HABITS AND HABITATS** Common *Calidris* wader found in freshwater wetlands such as rice fields, and muddy shores of ponds and lakes. **SITES** Candaba Marsh (Pampanga), Balanga (Bataan), rice fields and wetlands throughout the Philippines.

Curlew Sandpiper ■ *Calidris ferruginea* 20cm

DESCRIPTION Medium-small wader with characteristic decurved bill. Breeding birds dark rufous and grey all over, with face, neck, breast and belly rufous, and upperparts rufous mottled with grey. Non-breeding birds white and grey all over, with white face with dark grey eye-stripe, white eyebrow and dark crown. Throat white; underparts white with grey breast-band and white rump. Greyish-black legs. **DISTRIBUTION** Luzon, Mindoro, Cebu, Panay, Negros, Palawan, Mindanao. **HABITS AND HABITATS** Uncommon wader occurring in a wide variety of habitats, including shallow mudflats and flooded rice fields. **SITES** LPPCHEA (Metro Manila), Olango (Cebu), Balanga (Bataan).

Non-breeding

Oriental Pratincole ■ *Glareola maldivarum* 23cm

DESCRIPTION Medium-sized wader with very graceful flight. Breeding birds rufous-brown with lighter brown face, and cream throat bordered by thin black necklace. Upperparts darker reddish-brown; underparts lighter brown. Very short, thick black bill with a little orange at base of lower mandible. In flight deeply forked tail and white rump are diagnostic. **DISTRIBUTION** Luzon, Mindoro, Negros, Palawan. **HABITS AND HABITATS** Common resident wader, favouring open drier areas such as pastures and newly ploughed fields. Very graceful in flight, with similar flight patterns to terns. **SITES** Candaba Marsh (Pampanga).

LEFT & RIGHT: *Breeding adult*

Non-breeding

Black-headed Gull

■ *Chroicocephalus ridibundus* 38cm; WS 98.5cm

DESCRIPTION Medium-small gull with white head with small black spot behind eye. Rest of body all white with light grey back and wings, except outer primaries, which are white with black tips. In breeding plumage head turns brownish-black. Bill yellowish-orange tipped with black; legs reddish. **DISTRIBUTION** Luzon, Cebu, Mindanao, Mindoro, Palawan. **HABITS AND HABITATS** The most common gull in the Philippines, found mostly in coastal areas on tidal flats, bays and river deltas; sometimes also inland in flooded rice fields. Plucks prey from the water. **SITES** LPPCHEA (Metro Manila), Balanga (Bataan), Olango (Cebu).

Greater Crested Tern

Thalasseus bergii 43cm; WS 102cm

DESCRIPTION Large tern with distinctive short crest and long, decurved, all-yellow bill. Grey-and-white plumage with black crest and white forehead. Wings and back light grey. In flight deeply forked tail is noticeable. **DISTRIBUTION** Throughout the Philippines. **HABITS AND HABITATS** Common tern found along coast in bays and river deltas. Perches on bamboo poles of fish pens and sometimes on exposed mudflats. Dives for fish head first into the water. **SITES** LPPCHEA (Metro Manila), Balanga (Bataan), coastal areas throughout the Philippines.

Whiskered Tern

Chlidonias hybridus 25cm; WS 66cm

Breeding

DESCRIPTION Medium-small tern that is the most common tern in the Philippines. Shallow, forked tail, and white and light grey plumage. Breeding birds distinct, with black crown and red bill. Non-breeding birds have white forehead with dark grey patch at back of eye, grey upperparts and wings, and white underparts and rump. **DISTRIBUTION** Throughout the Philippines. **HABITS AND HABITATS** Common tern found in a wide variety of habitats, including coastal areas, river mouths, freshwater marshes, flooded rice fields and lakes. Hovers low above the water and dives head first for prey, unlike similar looking **White-winged Tern** *C. leucopterus*, which rarely dives. **SITES** LPPCHEA (Metro Manila), Balanga (Bataan), Candaba Marsh (Pampanga), Olango (Cebu), coastal areas and rice fields throughout the Philippines.

Red Turtle Dove *Streptopelia tranquebarica* 23cm

DESCRIPTION Medium-sized plump dove with grey head and narrow black collar, rest of body reddish-brown and short grey tail. Rump and outer tail white and edged with black. Male reddish; female duller greyish-brown. Black bill and legs. **DISTRIBUTION** Luzon, Mindoro, Mindanao. **HABITS AND HABITATS** Common ground-dwelling dove found in farms, open areas and mangroves. Forages on dry ground looking for fallen fruits and seeds. **SITES** Candaba Marsh (Pampanga), agricultural fields and around villages throughout the Philippines.

Male

Spotted Dove

Spilopelia chinensis 30cm

DESCRIPTION Medium-sized dove with grey head and diagnostic broad black collar with white spots. Throat light grey, and neck and breast light reddish-brown. Upperparts and wings darker brown. Long dark brown tail with white tips to outer feathers seen when flying or landing. Black bill, light orange eyes and red feet. **DISTRIBUTION** Throughout the Philippines. **HABITS AND HABITATS** Common dove occurring in open country, farms and forest edges. Ground feeder, foraging for fallen fruits and seeds. **SITES** Candaba Marsh (Pampanga), Sabang (Palawan), agricultural areas throughout the Philippines.

Adult

Common Emerald Dove ■ *Chalcophaps indica* 25cm

DESCRIPTION Small-medium dove with overall dark green and vinous plumage. Male bird has white forehead connecting with white stripe above eye, grey crown and nape. Throat, neck, breast and upper back pinkish-brown. Female more brown, with grey crown and pinkish-brown upperparts of male replaced by dull brown colour. Both male and female have diagnostic emerald-green wings and two broad silver-grey bands on black lower back and rump. Bright orange bill and red legs. **DISTRIBUTION** Throughout the Philippines. **HABITS AND HABITATS** Shy and secretive ground-dwelling dove found in lowland primary forests, secondary growth and forested parks. Feeds on the ground looking for fallen fruits, grubs and seeds. Flies off at the first sign of danger. **SITES** La Mesa Ecopark (Metro Manila), Subic Forest (Bataan), Mt Makiling (Laguna), PICOP (Surigao del Sur).

LEFT & RIGHT: *Male*

Zebra Dove

■ *Geopelia striata* 22cm

DESCRIPTION Small dove generally grey and brown in plumage. Grey head with brown crown, and bluish-grey cheeks and throat; neck and sides of breast light greyish-brown with characteristic black-and-white stripes. Rest of upperparts brown with edge of wing darker brown. Front of breast pinkish-brown; belly and vent white; long tail brown with white tips. **DISTRIBUTION** Throughout the Philippines. **HABITS AND HABITATS** The most common open-country dove, often seen walking on the ground foraging for fallen seeds and grubs. **SITES** La Mesa Ecopark (Metro Manila), UP Diliman (Metro Manila), fields and urban areas throughout most of the Philippines.

White-eared Brown Dove ■ *Phapitreron leucotis* 23cm e

DESCRIPTION Small, plump dove with overall brown plumage. Crown greyish-brown; black line starting from gape passing below eye and extending into nape, and white line or 'ear' below black line. Nape and collar greenish-bronze, upperparts darker reddish-brown and underparts lighter. Vent white and tail brown with grey tips. **DISTRIBUTION** Throughout the Philippines except Palawan. **HABITS AND HABITATS** The most common of the endemic doves of the Philippines, found in primary and secondary forests from lowlands up to 1,600m. Mostly seen in singles or pairs, but forms small flocks when feeding on fruiting tree. Feeds mainly on small fruits, including figs and berries. **SITES** Subic Forest (Bataan), Mt Makiling (Laguna), Mt Kitanglad (Bukidnon), PICOP (Surigao del Sur).

Adult, Luzon

Amethyst Brown Dove ■ *Phapitreron amethystinus* 27cm e

DESCRIPTION Superficially similar to White-eared Brown Dove (see above), but larger and more stocky, with stronger and heavier bill, and iridescent purple upper back and neck. Black line below eye also starts from gape and extends to nape; right below black line is white line of same length. Undertail tinged with rufous. **DISTRIBUTION** Luzon, Samar, Leyte, Bohol, Negros, Cebu, Mindanao. **HABITS AND HABITATS** The largest of the brown doves in the Philippines, and locally common in some areas of primary and secondary forests. Favours middle to high elevations, but occurs from lowlands to 2,000m. Usually in singles or pairs, but also flocks in fruiting fig trees. **SITES** Subic Forest (Bataan), Mt Makiling (Laguna), Mt Kitanglad (Bukidnon), PICOP (Surigao del Sur).

Adult, Luzon

Philippine Green Pigeon ■ *Treron axillaris* 28cm e

DESCRIPTION Medium-large pigeon with overall green-and-black plumage. Small light grey crown; rest of body yellowish-green. Male has maroon mantle and black wings with two yellow bars. Female lacks maroon mantle; colour replaced with darker yellowish-green. Bluish bill with red base. The only green pigeon in the Philippines with light blue eyes. **DISTRIBUTION** Luzon, Mindoro, Cebu, Bohol, Samar, Leyte, Mindanao, Negros, Panay. **HABITS AND HABITATS** Uncommon pigeon found in primary and secondary growth forests from lowlands up to 1,000m. Often seen in small flocks, but forms larger flocks when feeding in fruiting trees, often together with other doves. **SITES** Subic Forest (Bataan), Rajah Sikatuna National Park (Bohol), PICOP (Surigao del Sur).

Male, Luzon

Flame-breasted Fruit Dove ■ *Ptilinopus marchei* 38cm e

DESCRIPTION Very large and distinctive fruit dove. Crimson-red crown and nape and black face-patch. Red bill; small white ring surrounding red eye. Upperparts and wings blackish glossy green with red wing-patch. Underparts grey with very conspicuous bright orange throat and red upper breast. Red legs. **DISTRIBUTION** Luzon. **HABITS AND HABITATS** Uncommon pigeon that is the largest of the fruit doves in the Philippines. Only found in high-elevation montane mossy forests, generally above 1,000m. Very shy and hard to see, and will often fly off before being seen, departing with loud wingbeats. **SITES** Mt Polis (Mountain Province), Sierra Madre (Cagayan), Mt Banahaw (Quezon).

Adult

Cream-breasted Fruit Dove ■ *Ptilinopus merrilli* 32cm ⓔ

DESCRIPTION Medium-large pigeon with drab grey-and-green plumage. Grey head with bright orange eyes and bill, and white throat. Upperparts and wings dull green, with black wing-tips and chestnut underwings. Breast grey-green, contrasting with buff-cream belly, which gives species its name. Green flanks, lower belly and tail. Red legs. The *faustinoi* race from Northern Luzon has bright red cap. **DISTRIBUTION** Luzon. **HABITS AND HABITATS** Uncommon pigeon favouring the canopy of lowland primary and secondary forests. Very shy and has seasonal local movements depending on presence of fruiting trees. **SITES** Sierra Madre (Cagayan).

Adult, Northern Luzon

Yellow-breasted Fruit Dove ■ *Ptilinopus occipitalis* 30cm ⓔ

DESCRIPTION Medium-sized fruit dove that is one of the Philippines' most colourful doves. Grey crown and white forehead bordered by maroon on sides of face and on nape. Upperparts green and upper breast white, accented with big yellow breast-patch. Broad maroon band on upper belly bordering greyish-green lower belly. Orange bill with yellow tip. Red legs. **DISTRIBUTION** Throughout the Philippines except Palawan. **HABITS AND HABITATS** The most common fruit dove in the country, found in the canopies of lowland primary and secondary forests up to 1,800m. Feeds mainly on fruits, and flocks in fruiting trees. **SITES** Subic Forest (Bataan), Rajah Sikatuna National Park (Bohol), Mt Kitanglad (Bukidnon).

Adult, Mindanao

Black-chinned Fruit Dove ■ *Ptilinopus leclancheri* 25–28cm

DESCRIPTION Medium-sized pigeon with pale grey head, neck and breast, red eyes and yellow bill. Upperparts all-drab green; underparts lighter olive-green bordered by dark maroon breast-band, rufous-brown vent and tail tipped with grey. Female similar to male, with grey face, neck and breast replaced by green. Both sexes have small black chins. Feet are red. **DISTRIBUTION** Throughout the Philippines. **HABITS AND HABITATS** Uncommon pigeon found in lowland forests. Stays up in the canopy foraging for fruits. **SITES** Mt Makiling (Laguna), Sabang (Palawan), PICOP (Surigao del Sur).

Adult, Palawan

Green Imperial Pigeon ■ *Ducula aenea* 45cm

DESCRIPTION Large pigeon with light pinkish-grey head, breast and belly, and green back, wings and tail. Orange eyes and grey bill. Undertail-coverts deep chestnut. **DISTRIBUTION** Throughout the Philippines. **HABITS AND HABITATS** Common pigeon that favours lowland forests and loves to perch on dead branches of the tallest trees in the early morning. Feeds on large fruits, sometimes with other pigeons or hornbills. Also seen in mangroves. **SITES** Subic Forest (Bataan), Puerto Princesa Underground River Subterranean Park (Palawan), Rajah Sikatuna National Park (Bohol).

Adult, Southern Luzon

Grey Imperial Pigeon

■ *Ducula pickeringii* 41cm

DESCRIPTION Large imperial pigeon that is very similar to Green Imperial Pigeon (see p. 55), from which it is distinguished by its brownish-grey wings and pinkish-grey rather than chestnut undertail-coverts. **DISTRIBUTION** Small offshore islands in Palawan and Sulu. **HABITS AND HABITATS** Locally common, highly specialized pigeon preferring to live on small islands. Feeds mainly on fruits and will fly from one island to the next in search of fruiting trees. Has also been observed to eat leaves, presumably when fruits are not in season in its range. **SITES** Rasa Island (Palawan), Honda Bay (Palawan).

Pied Imperial Pigeon ■ *Ducula bicolor* 38cm

DESCRIPTION Large pigeon with black-and-white plumage. Entire body creamy-white except for black flight feathers and tail. Brown eyes, metallic bluish-grey bill with black tip, and grey legs. **DISTRIBUTION** Offshore islands throughout the Philippines. **HABITS AND HABITATS** Locally common imperial pigeon, roosting and nesting on small offshore islands. During the day flies in big groups, visiting other small islands and the mainland to feed. **SITES** Rasa Island (Palawan), Honda Bay (Palawan).

Red-vented Cockatoo

■ *Cacatua haematuropygia* 31cm ⓔ

DESCRIPTION Medium-sized parrot with almost wholly cream-white plumage. Small white crest sometimes with tinge of yellow, greyish bill and dark brown eyes surrounded by light blue skin. Underwing has varying amounts of yellow, and undertail-coverts and vent are reddish-orange with all-yellow undertail. Strong, powerful, bluish-grey legs. **DISTRIBUTION** Formerly throughout the Philippines. Population stronghold now on Palawan, although still present in small numbers on Bohol, Samar and Panay. **HABITS AND HABITATS** The Philippines' only cockatoo, and now very rare in lowland, riverine, mangrove and coastal forests. Very social bird, sometimes seen in singles or pairs. Mostly roosts in family groups in nest holes, dead snags of tall mangroves and even coconut groves. **SITES** Rasa Island (Palawan), Sabang (Palawan). **CONSERVATION** Severely threatened by rampant poaching for the illegal pet trade and considered Critically Endangered. Disappearance of suitable mangrove and coastal forest due to habitat destruction and over-development also contribute to its decline.

Philippine Hanging Parrot/Colasisi

■ *Loriculus philippensis* 15cm ⓔ

DESCRIPTION Small parrot with generally green plumage with patches of red. Several races in the Philippines, which differ considerably. The *philippensis* race that occurs in Luzon has bright red forehead, and green face with varying amounts of blue. Rest of body all green except for light cobalt-blue underwing and undertail, and bright red rump and uppertail-coverts. Male has bright red breast-patch. Bright red bill and legs. Other subspecies differ by amount of red on forehead and crown. **DISTRIBUTION** Throughout the Philippines except Palawan. **HABITS AND HABITATS** The smallest parrot in the Philippines, commonly found in all forest types, and even in green patches and gardens in cities. Feeds mainly on fruits, flowers and coconut blossoms. High-pitched call usually given while on its fast, undulating flight. **SITES** UP Diliman (Metro Manila), Subic Forest (Bataan), Mt Kitanglad (Bukidnon).

Female, Mindanao

Green Racket-tail ■ *Prioniturus luconensis* 22cm (e)

DESCRIPTION Small-medium parrot with all-green plumage. Head, neck and underparts yellow-green; rest of body darker yellow-green. Adult birds have two thin, wire-like black extensions on tail ending in blue-black spatules that look like tennis racquets. Metallic grey bill and legs. **DISTRIBUTION** Luzon. **HABITS AND HABITATS** Uncommon parrot that is the smallest of the seven endemic racket-tails in the Philippines. Found in lowland forests and forest edges up to 1,000m. Green plumage blends well in forest canopy, but bird's presence is given away by raucous shrieking calls. **SITES** Subic Forest (Bataan). **CONSERVATION** Endangered parrot that is continually declining and is now regularly recorded only in Subic Forest.

Adult, Luzon

Blue-naped Parrot

■ *Tanygnathus lucionensis* 32cm

DESCRIPTION Medium-sized parrot with mostly yellowish-green plumage. Bright green head with blue hind-crown and nape. Upperparts yellowish-green; wings a mix of green-and-blue feathers. Bright red bill and grey feet. **DISTRIBUTION** Formerly all over the Philippines, but now only regularly seen in Subic Forest and Palawan. **HABITS AND HABITATS** Very noisy parrot occurring in forests, forest edges and mangrove forests. Feeds on fruits and flowers, usually in pairs or small groups. Uses holes made by woodpeckers for nesting. Heavily trapped for the illegal pet trade. **SITES** Subic Forest (Bataan), Puerto Princesa Underground River Subterranean Park (Palawan).

Guaiabero ■ *Bolbopsittacus lunulatus* 17cm ⓔ

DESCRIPTION Small, plump parrot that is all green and has a very short tail. Male has light blue wash on face, and very thin blue line forming a necklace. Female has less blue on face, usually just around eye, and has black-and-yellow necklace. Silver-grey bill, and grey legs and feet. **DISTRIBUTION** Luzon, Samar, Leyte, Mindanao. **HABITS AND HABITATS** Found primarily in lowland forests and forest edges, and sometimes even in green patches in the city. Prefers to feed on fruiting figs and other fruits, especially guava. Usually in pairs or small groups. Fast, direct flight; all-green plumage blends well with foliage. **SITES** Subic Forest (Bataan), Mt Makiling (Laguna), PICOP (Surigao del Sur).

Male, Luzon

Black-faced Coucal

■ *Centropus melanops* 48cm ⓔ

DESCRIPTION Large coucal with yellowish-buff head, neck, throat, mantle and upper breast accented by black face. Wings and rest of upperparts chestnut; lower breast brownish-black, becoming darker in undertail-coverts. Long tail black with bluish gloss. Deep red eyes and black bill. **DISTRIBUTION** Samar, Leyte, Bohol, Mindanao. **HABITS AND HABITATS** Uncommon inhabitant of lowland forests and forest edges. Found singly or in pairs, often skulking in dense tangles of middle and upper storeys of forests. Like most coucals, a weak and clumsy flyer. Loud, four-note, booming call similar to hoots of fruit doves. **SITES** PICOP Rajah Sikatuna National Park (Bohol), PICOP (Surigao del Sur).

Philippine Coucal ■ *Centropus viridis* 42cm ⓔ

DESCRIPTION Medium-sized coucal that is the most common coucal in the Philippines. Black all over with chestnut wings. Deep red eyes, and black bill and legs. The *mindorensis* race, occurring only in Mindoro, is all black, including wings. Immature birds similar to adults, but with head and wings having fine chestnut-and-black barring. **DISTRIBUTION** Throughout the Philippines except Palawan. **HABITS AND HABITATS** Common coucal found in a variety of habitats, including wooded grassland, bamboo groves, and lowland and montane forests up to 2,000m. Occurs singly or in pairs, and like most coucals is very shy and secretive. Best time to see it is after rain, when birds are drying out their weak, stubby wings. **SITES** Subic Forest (Bataan), Mt Kitanglad (Bukidnon).

Lesser Coucal

■ *Centropus bengalensis* 36cm

DESCRIPTION Smaller than similar Philippine Coucal (see above). Black all over with bright buff shafts in scapulars, and dark brown- and rufous-mottled wings. Immature birds are distinctive: light brown with buff shafts in upperparts and finely barred tail. **DISTRIBUTION** Throughout the Philippines. **HABITS AND HABITATS** Common inhabitant of open areas and grassland. A skulker, hopping and moving through dense foliage, but sometimes perches on tops of grass patches while calling. **SITES** Candaba Marsh (Pampanga), UP Los Baños (Laguna).

Chestnut-breasted Malkoha ■ *Phaenicophaeus curvirostris* 46cm

DESCRIPTION Large malkoha with bright lime-green bill, bare red face and dark olive-green crown extending up to upper back. Sides of neck and upper breast chestnut; a little darker in rest of underparts. Wings emerald-green. Long tail two toned, with first half dark green, becoming chestnut in tip. Eyes red in males and yellow in females; bound by whitish eye-ring in both sexes. **DISTRIBUTION** Palawan. **HABITS AND HABITATS** Common malkoha that favours lowland forests and forest edges. Often in pairs, and sometimes in family groups. A skulker, climbing and moving about in dense foliage. **SITES** Sabang (Palawan).

Scale-feathered Malkoha

■ *Dasylophus cumingi* 41cm e

DESCRIPTION Large malkoha with unique head markings. Light grey head with plastic-like black scales on crown running up to nape, and on centre of throat running down to breast. Red eyes surrounded by bare red skin. Pale yellow bill. Back and breast chestnut and wings black with greenish gloss. Long tail black with greenish gloss and white tip. **DISTRIBUTION** Luzon. **HABITS AND HABITATS** Common malkoha of lowland and montane forests. Usually in pairs or small flocks, skulking in dense tangles and vines of forest understorey. More common in middle to high elevations than **Rough-crested Malkoha** *D. superciliosus*. **SITES** Subic Forest (Bataan), Mt Makiling (Laguna), Mt Polis (Mountain Province).

Rusty-breasted Cuckoo

■ *Cacomantis sepulcralis* 22cm

DESCRIPTION Small cuckoo very similar to **Plaintive Cuckoo** C. *merulinus*. Dark grey crown, and lighter grey face and throat. Upperparts and wings greyish-brown, with breast, belly and vent rufous. Tail bluish-black with white tips. Brown eyes with yellow eye-ring. Two-toned bill, with upper bill black and lower bill yellow. Yellow legs. Immature birds all rufous, with black barring in both upperparts and underparts. **DISTRIBUTION** Throughout the Philippines. **HABITS AND HABITATS** Common cuckoo found in mangoves, forests and forest edges from lowlands up to montane forests up to 2,000m. Known nest parasite. Loud, ascending call usually given when perched in the canopy. Previously lumped with widespread **Brush Cuckoo** C. *variolosus*. **SITES** Mt Makiling (Laguna), Mt Kitanglad (Bukidnon).

Philippine Drongo Cuckoo ■ *Surniculus velutinus* 23cm e

DESCRIPTION Size of small cuckoo, with all-glossy black plumage. Slim build and small, thin, decurved bill that separates it from true drongos. Slightly forked tail.

DISTRIBUTION Throughout the Philippines except Palawan. **HABITS AND HABITATS** Quite common and found in lowland forests and forest edges, usually perched in the canopy and middle storey. Secretive, but has loud, conspicuous call, given when perched. **SITES** Mt Makiling (Laguna), Rajah Sikatuna National Park (Bohol).

Philippine Hawk-Cuckoo ■ *Hierococcyx pectoralis* 29cm ⓔ

DESCRIPTION Medium-sized cuckoo with light grey head, rufous breast and belly, and white underparts. Back and wings darker grey. Tail long with 3–4 dark bands and black terminal band with rufous tip. Eyes reddish-brown surrounded by yellow eye-ring. Upper bill black and lower bill yellow. Light orange legs. **DISTRIBUTION** Throughout the Philippines. **HABITS AND HABITATS** Uncommon endemic cuckoo that resides in lowlands up to montane forests and secondary growth. Resembles small *Accipiter* and forages in the canopy and middle storey, looking for caterpillars and other insects. Secretive and perches motionless for long periods. **SITES** Mt Makiling (Laguna), Mt Kitanglad (Bukidnon).

Immature

Eastern Grass Owl ■ *Tyto longimembris* 40cm

DESCRIPTION Large owl with ghostly-white plumage and distinctive heart-shaped facial pattern. Round white face, no ear-tufts, light grey crown, white underparts, and golden-brown upperparts and wings mottled with white spots. Barred tail and flight feathers. **DISTRIBUTION** Throughout the Philippines. **HABITS AND HABITATS** Nocturnal and prefers grassland and rice fields in open country. Hunts low over fields near twilight, foraging for field mice, lizards and amphibians. Roosts in dense, tall grass during the day. **SITES** Candaba Marsh (Pampanga), Bislig Airfield (Surigao del Sur).

Giant Scops Owl

■ *Otus gurneyi* 31cm e

DESCRIPTION Medium-sized owl with long ear-tufts. Rufous facial disc with white eye-stripes from lores up to horns, forming 'V'. Upperparts darker rufous with broad black striations. Underparts lighter brown. Light rufous throat forms necklace. **DISTRIBUTION** Mindanao, Samar. **HABITS AND HABITATS** Common, strictly nocturnal owl found from lowlands up to mountains in forests and forest edges, and sometimes in wooded developed areas. Shy and difficult to see, with an explosive call that is given irregularly. **SITES** Mt Kitanglad (Bukidnon), PICOP (Surigao del Sur).

Palawan Scops Owl

■ *Otus fuliginosus* 20cm e

DESCRIPTION Small owl with ear-tufts. Rufous facial disc accented by buff-white eye-stripes running from lores up to ear-tufts, forming 'V'. Forehead dark brown; rest of face is a little lighter. Distinctive buff collar; upperparts darker with barring and underparts slightly lighter. **DISTRIBUTION** Palawan. **HABITS AND HABITATS** Uncommon owl that is the only small-eared owl in Palawan. Found in forests and forest edges in dense undergrowth and bamboo groves. Quiet growling call. **SITES** Sabang (Palawan).

Philippine Scops Owl

■ *Otus megalotis* 28cm ⓔ

DESCRIPTION Medium-sized owl with light brown eye-stripes from lores to ear-tufts, forming 'V'. Forehead dark brown; distinct dark brown collar. Upperparts darker mottled brown; underparts lighter with dark striations. Red eyes and ivory-horn bill and cere. Feathered legs with grey feet. **DISTRIBUTION** Luzon. **HABITS AND HABITATS** Common owl that prefers understorey of forests and forest edges, sometimes in urban areas like gardens with lots of trees, from lowlands up to high-elevation mountains. Largest of the Philippine *Otus* owls. Feeds on a variety of prey, including mice, lizards, geckos and even small birds. **SITES** Subic Forest (Bataan), Mt Makiling (Laguna).

Everett's Scops Owl

■ *Otus everetti* 22cm ⓔ

DESCRIPTION This small owl used to be lumped with the Philippine Scops Owl (see above), from which it differs in its smaller size and having a darker blackish-brown crown and face. It has a mottled tan eye-stripe, dark ear-tufts, red eyes and light brown bill. **DISTRIBUTION** Mindanao, Bohol. **HABITS AND HABITATS** Uncommon owl found in forests and forest clearings from lowlands up to mid-montane regions. Prefers to hunt low in the understorey, and sometimes goes to the ground to catch lizards and insects. **SITES** Rajah Sikatuna National Park (Bohol), PICOP (Surigao del Sur), Mt Kitanglad (Bukidnon).

Negros Scops Owl

■ *Otus nigrorum* 20cm ⓔ

DESCRIPTION This is another split from the Philippine Scops Owl (see p. 65). Small owl with rufous facial disc, short ear-tufts and bright white eye-stripe starting from bill, up to rufous ear-tufts. Eyes red and bill ivory-horn tipped with black. Nape and collar rufous; upperparts and wings mottled dark brown. Underparts lighter brown and buff-white with striations. **DISTRIBUTION** Negros and Panay. **HABITS AND HABITATS** Uncommon and lives in primary forests, secondary growth and/ or wooded gardens. Hunts for small rodents and lizards, and nests in tree cavities. **SITES** Twin Lakes (Negros), Forest Camp, Valencia (Negros).

Luzon Scops Owl

■ *Otus longicornis* 18cm ⓔ

DESCRIPTION Tiny owl with overall dark brown mottled plumage. Whitish lores, yellow eyes and buff ear-tufts. Bill two toned, with upper bill darker yellowish-horn than lower bill. Upperparts and wings dark mottled brown; underparts lighter. **DISTRIBUTION** Luzon. **HABITS AND HABITATS** Locally common owl confined to high-elevation mountains of Luzon, in mossy and pine forests above 1,000m. Utters a nice, soft, two-note call, given regularly. Nests in tree hollows. **SITES** Mt Polis (Mountain Province).

Mantanani Scops Owl

■ *Otus mantananensis* 23cm

DESCRIPTION Medium-sized owl with rufous facial disc, short ear-tufts, black whiskers that form dark moustache, yellow eyes and dirty-brown bill. Upperparts reddish-brown with very fine streaks of dark brown. Wings with buff-white barring forming buff bar on scapulars. Rest of underparts rufous with streaks of dark brown and black running down breast. **DISTRIBUTION** Islands off mainland Palawan, Romblon, Tablas, Sibuyan, Semirara, small islands in the Sulu Archipelago. **HABITS AND HABITATS** Locally common owl that is a small-island specialist, preferring to live on small offshore islands, never on big islands. Lives in a wide variety of habitats, including mangrove forests and wooded areas, and even coconut plantations. A near-endemic owl with small outlying populations on offshore islands of Borneo. **SITES** Rasa Island (Palawan), Pandan Island (Palawan), Dubduban Watershed (Tablas).

Philippine Eagle-owl

■ *Bubo philippensis* 51cm ⓔ

DESCRIPTION Large, stocky owl with light rufous facial disc, rufous ear-tufts, rufous crown with darker brown streaks, yellow eyes and ivory-horn bill. White eye-stripe from buff-white lores up to above eyes. Upperparts rufous with broad streaks; underparts lighter brown with broad dark brown streaks. Powerful legs fleshy-brown. **DISTRIBUTION** Luzon, Catanduanes, Bohol, Leyte, Samar, Mindanao. **HABITS AND HABITATS** Uncommon owl that is the largest owl in the Philippines. Found in primary and secondary forests, but sometimes also in developed areas with suitable nesting grounds and availability of prey. Primarily hunts vertebrate prey such as rodents, amphibians and small birds. **SITES** Angono Petroglyphs (Rizal), Subic Forest (Bataan).

Immature, Luzon

Spotted Wood Owl

■ *Strix seloputo* 44cm

DESCRIPTION Large owl with very round head and rufous facial disc. No ear-tufts and very dark brown eyes. Chestnut crown and dark brown lores connecting to crown divide facial disc. Upperparts all-dark brown mottled with white spots; throat white and rest of underparts lighter rufous-brown with very fine barring. Black bill and cere. Strong, powerful yellow claws are feathered. **DISTRIBUTION** Mainland Palawan. **HABITS AND HABITATS** Uncommon, poorly known owl that is the only large owl found in Palawan. Inhabits lowland forests and secondary growth. Prefers to roost high up in the canopy and has a deep, far-reaching, hooting call. **SITES** Sabang (Palawan).

Chocolate Boobook

■ *Ninox randi* 31cm

DESCRIPTION Medium-large owl with no ear-tufts. Dark brown head with white forehead, lores and chin. Yellow eyes and dark grey bill. Upperparts uniform dark brown; underparts lighter chocolate-brown with broad white bands running from upper breast to belly. Tail has alternating dark and less dark bands. Yellow feet. **DISTRIBUTION** Luzon, Mindanao, Negros, Mindoro, Cebu. **HABITS AND HABITATS** Uncommon owl that used to be included in the Philippine Hawk-Owl complex. Inhabits primary and secondary forests from lowlands to higher elevations. **SITES** Subic Forest (Bataan), PICOP (Surigao del Sur).

Luzon Hawk-Owl

Ninox philippensis 21cm e

DESCRIPTION Small-medium owl with rounded head and no ear-tufts. White lores and moustache that forms white 'X' around yellow bill. Yellow eyes. Head darker brown with buff-white chin. Upperparts brown; underparts buff-white streaked heavily with broad brown stripes. Unfeathered yellow legs. **DISTRIBUTION** Luzon, Samar, Leyte, Negros, Panay, Bohol, Masbate, Ticao, Siquijor. **HABITS AND HABITATS** Common earless owl found in primary and secondary lowland forests. It used to be part of a bigger complex of Philippine Hawk-Owls, but recent studies have split the group into several distinct species. Feeds on a variety of prey such as small lizards, rodents and insects. **SITES** Subic Forest (Bataan), Mt Makiling (Laguna), Rajah Sikatuna National Park (Bohol).

Mindanao Hawk-Owl

Ninox spilocephala 21cm e

DESCRIPTION Small-medium, earless owl with very rounded head. Very similar to Luzon Hawk-Owl (see above), with yellow eyes and white 'X' pattern on face formed by white supercilium. Differs from Luzon Hawk-Owl by having spotted dark brown crown and nape. Upperparts brown; underparts buff with broad brown stripes like those of Luzon Hawk-Owl, but less defined. **DISTRIBUTION** Mindanao, Dinagat, Siargao, Basilan. **HABITS AND HABITATS** Uncommon owl that is Near Threatened due to continued decline of its preferred habitat of lowland primary and secondary forests. Used to belong to Philippine Hawk-Owl complex, but due to major difference in its calls and body measurements has been elevated to full species status. **SITES** PICOP (Surigao del Sur).

Mindoro Hawk-Owl ■ *Ninox mindorensis* 21cm e

DESCRIPTION Small-medium owl with no ear-tufts and very rounded head. Finely barred crown with short white supercilium forming 'X' pattern on face. Eyes yellow and throat buff-white, forming distinct throat-patch. Upperparts darker brown; underparts lighter with very fine horizontal barring. **DISTRIBUTION** Mindoro. **HABITS AND HABITATS** Uncommon owl that prefers primary and secondary forests and forest edges. Usually found singly or in pairs in lowlands of Mindoro. Used to be lumped with the other hawk-owls in the Philippines, but due to its distinct call and fine horizontal barring (as opposed to broad downwards stripes) on underparts has been elevated to full species status. **SITES** Sablayan (Mindoro).

Cebu Hawk-Owl ■ *Ninox rumseyi* 27cm e

DESCRIPTION Medium-sized owl with big, rounded head and no ear-tufts. Crown dark brown with dark speckles, eyes yellow with short white supercilia, and distinct white throat-patch. Upperparts dark brown with darker spots on wings and white patch near scapulars. Underparts uniform rufous-brown with barring on chest, barring being highly variable between individuals. Whitish vent and yellow legs. **DISTRIBUTION** Cebu. **HABITS AND HABITATS** Range-restricted owl found in remaining lowland forests of Cebu. One of the largest of the Philippine Hawk-Owl complex, and preys on a variety of vertebrates as well as insects. Nests in tree hollows. Elevated to full species status in mid-2012 from Philippine hawk-owl complex. **SITES** Tabunan (Cebu), Alcoy (Cebu).

Camiguin Hawk-Owl ■ *Ninox leventisi* 27cm ⓔ

DESCRIPTION Medium-sized hawk-owl with rounded head and no ear-tufts; similar in appearance and size to Cebu Hawk-Owl (see opposite). Dark brown barred crown. Lacks white supercilium present in other hawk-owls; white throat-patch and distinctive grey-blue eyes. Dark brown upperparts with white patch in scapulars; dark brown underparts with consistent heavy barring. **DISTRIBUTION** Camiguin Sur. **HABITS AND HABITATS** Uncommon owl that can be found in primary and secondary lowland forests of Camiguin. The only owl in the Philippines with bluish-grey eyes. Usually found singly or in pairs. **SITES** Camiguin Sur. **CONSERVATION** Due to continuing habitat degradation and loss in a very restricted range, population is believed to be in continuing decline; species classified as Endangered.

Philippine Frogmouth ■ *Batrachostomus septimus* 23cm ⓔ

DESCRIPTION Medium-small frogmouth. Brown all over with broad bill with very wide gape resembling that of a frog. Lots of whiskers emanate from lores and sides of face. Deep red eyes. Occurs in two phases: rufous and brown. Birds in brown phase have mottled brown upperparts; lighter brown underparts with two broad buff-white bands in upper breast and lower breast. In rufous-phase plumage less mottled and browns are replaced by rufous. **DISTRIBUTION** Luzon, Mindanao, Samar, Leyte, Bohol, Negros, Panay. **HABITS AND HABITATS** Uncommon, strictly nocturnal bird found in both lowland and high-elevation forests and forest edges. During the day perched upright, mimicking dried, broken branch. **SITES** Mt Kitanglad (Bukidnon), PICOP (Surigao del Sur).

Palawan Frogmouth

■ *Batrachostomus chaseni* 21cm e

DESCRIPTION Small frogmouth similar to Philippine Frogmouth (see p. 71) in plumage, but has a very different call. Rufous-brown face with ear-tufts, and long rufous whiskers emanating from lores and sides of head. Yellow eyes. Upperparts dark rufous-brown with some white spots on scapulars; underparts lighter rufous-brown with buff band on upper breast and large buff-white patch on belly. **DISTRIBUTION** Palawan. **HABITS AND HABITATS** Common in primary and secondary forests and forest edges in lowland Palawan. Wide range of calls, from soft whistle to harsh growl. **SITES** Sabang (Palawan).

Great Eared Nightjar

■ *Lyncornis macrotis* 35cm

DESCRIPTION Large nightjar with long wings, long tail and generally all-brown plumage. When perched, ear-tufts very visible. Chocolate-brown head with dark brown spots on forehead and nape; ear-tufts lighter brown; white throat connecting with light chocolate-brown collar. Upperparts dark brown mottled with chocolate-brown and dark brown spots. Underparts also dark brown, with similar mottling and lighter brown upper breast-band. In flight no white on wings is visible, and tail has alternating light and dark brown bands without any white spots. **DISTRIBUTION** Luzon, Mindoro, Samar, Leyte, Bohol, Mindanao. **HABITS AND HABITATS** Common nightjar that is the largest in the Philippines. Very active at dawn and dusk, flying over forests and forest edges with *Accipiter*-like flight. Found from lowlands up to montane areas. Also perches in tall, dead trees at night. During the day remains hidden, sitting on the ground, using plumage as camouflage. **SITES** Subic Forest (Bataan), Mt Kitanglad (Bukidnon), PICOP (Surigao del Sur).

Large-tailed Nightjar

Caprimulgus macrurus 26cm

DESCRIPTION Medium-sized nightjar. All brown with upperparts streaked with black stripes. Darker chocolate-brown face and white throat connecting to lighter brown collar. Underparts brown, heavily streaked with black. In flight wings reveal white spots near tips, and tail has broad white terminal spots. **DISTRIBUTION** Palawan. **HABITS AND HABITATS** Common nightjar found in open country, mangroves and forest edges in lowland Palawan. Often seen on the ground in the middle of unpaved roads at night. Hunts for insects at night; during the day remains on the ground hidden among dried vegetation. **SITES** Sabang (Palawan), Iwahig (Palawan).

Philippine Nightjar

Caprimulgus manillensis 24cm e

DESCRIPTION Medium-sized nightjar very similar to Large-tailed Nightjar (see above) but smaller. Upperparts and underparts brown, heavily streaked with black; broken white throat. In flight shows buff-white spots near wings-tips and two terminal white spots on tail. **DISTRIBUTION** Throughout the Philippines except Palawan. **HABITS AND HABITATS** Uncommon nightjar that prefers forests and forest edges, scrub, bamboo groves, coconut plantations and even wooded urban areas. Very active at dawn and dusk, hawking for insects from favourite perches. Sits on coconut fronds and other prominent objects while giving *chuk-chur* call. **SITES** Mt Makiling (Laguna), UP Diliman (Metro Manila), Mt Kitanglad (Bukidnon), PICOP (Surigao del Sur).

Savanna Nightjar ■ *Caprimulgus affinis* 21cm

DESCRIPTION Small-medium nightjar that is grey-brown all over, with upperparts and underparts mottled grey, white and black. Short, thin white eyebrow. Like Philippine Nightjar (see p. 73), has a broken white throat. In flight shows white spots near wing-tips, and two rows of white bands from end of vent up to tail-tip. **DISTRIBUTION** Luzon, Mindoro, Negros, Panay, Mindanao. **HABITS AND HABITATS** Uncommon nightjar that is the smallest in the Philippines. Favours undisturbed beaches, sandy shores near mangrove areas, dry agricultural fields and rocky river beds. Crepuscular, flying and hawking for insects on the wing. During daytime grey plumage blends well in its sandy habitat. **SITES** Laoag Rivermouth (Ilocos Norte), Sablayan (Mindoro).

Whiskered Treeswift ■ *Hemiprocne comata* 16cm

DESCRIPTION Medium-sized treeswift with dark metallic-blue head, accented by unique white 'whiskers' above and below eye, extending past nape. Male has chestnut ear-patch. Long, dark metallic-blue wings extend way past tail. Back and upper belly brown-olive; lower belly and vent white. **DISTRIBUTION** Throughout the Philippines except Palawan. **HABITS AND HABITATS** Fairly common bird found in forests and forest edges from lowlands to high-elevation mountains. Prefers to perch on exposed leafless branches, where it scouts for prey. Distinct behaviour of flying out from perch and catching insects on the wing before returning to the same spot. **SITES** Subic Forest (Bataan), Mt Kitanglad (Bukidnon), PICOP (Surigao del Sur).

Grey-rumped Swiftlet ■ *Collocalia marginata* 9.4cm ⓔ

DESCRIPTION The smallest Philippine swiftlet, with glossy blue upperparts, white belly and slightly forked tail. Some races (*marginata* and *septentrionalis*) have off-white greyish rump, while others (*bagobo*) do not have grey rump. All races have grey face and chin, tiny black bill and dark brown eyes. **DISTRIBUTION** Throughout the Philippines. **HABITS AND HABITATS** The most common swiftlet in the Philippines. Can be found in a wide variety of habitats, from coastal areas up to high-elevation mountains. Feeds on insects on the wing. Nests on cliffs and walls of buildings, and at the entrances of caves, using moss and saliva to make nest. **SITES** Mt Makiling (Laguna), in various habitats throughout the Philippines.

Adult, Luzon

Ameline Swiftlet ■ *Aerodramus amelis* 12.2cm ⓔ

DESCRIPTION Medium-sized swiftlet with slightly forked tail. Blackish-brown upperparts and greyish-brown underparts. **DISTRIBUTION** Luzon, Cebu, Mindanao, Mindoro, Bohol, Batan. **HABITS AND HABITATS** Common swiftlet that was previously lumped with **Uniform Swiftlet** A. *vanikorensis*. Can be found in small groups in a wide variety of habitats, including grassland, forest, mangroves and coastal areas. Generally occurs below 1,000m. **SITES** Subic Forest (Bataan), PICOP (Surigao del Sur).

Philippine Spine-tailed Swift ■ *Mearnsia picina* 12cm e

DESCRIPTION Medium-large swift with all-black, glossy blue plumage. Relatively short body compared with its long 'butter knife'-shaped wings. Diagnostic white throat and white patches on underwing-coverts. Black tail has short extensions or 'spines'. **DISTRIBUTION** Mindanao, Negros, Samar, Leyte, Cebu. **HABITS AND HABITATS** Locally common swift favouring open areas adjacent to lowland forests. Fast and powerful flyer, usually in small groups or squadrons hunting for insects on the fly. **SITES** Mt Talinis (Negros), PICOP (Surigao del Sur), Baluno (Zamboanga).

Brown-backed Needletail ■ *Hirundapus giganteus* 22cm

DESCRIPTION Large and stocky needletail. Glossy black upperparts with greyish-brown back and rump. Underparts greyish-brown with white upper flanks extending to white undertail to form distinctive 'horseshoe' shape. Tip of tail greyish-brown with small spines or 'needles'. **DISTRIBUTION** Palawan. **HABITS AND HABITATS** Uncommon needletail usually seen flying fast over forests or open areas. Powerful yet agile with great manoeuvrability, catching insects on the wing. Tiny legs that are weak and rarely used as the birds spend most of their lives in the air. **SITES** Sabang (Palawan).

Purple Needletail ■ *Hirundapus celebensis* 24cm

DESCRIPTION Large needletail very similar in appearance to Brown-backed Needletail (see opposite), with which its range does not overlap. Mostly black plumage with bluish gloss. Underparts brown-black with white undertail-coverts connecting to white flanks, forming 'horseshoe'. Distinguished from Brown-backed by having white lores. Like other needletails, has spines at tip of black tail. **DISTRIBUTION** Luzon, Mindanao, Mindoro, Negros, Panay, Leyte. **HABITS AND HABITATS** Extremely fast flyer. Quite common over forest clearings and mountain valleys, and sometimes even in urban areas like Manila. When flying low and close, sound of the wind swooshing can be heard due to its fast and powerful flight. **SITES** Subic Forest (Bataan), Sierra Madre (Cagayan), PICOP (Surigao del Sur).

Philippine Trogon

■ *Harpactes ardens* 30cm (e)

DESCRIPTION Medium-sized, colourful bird with stocky body and long tail. Male has black forehead and throat, and maroon nape. Face has bare cobalt-blue skin surrounding dark brown eye. Bill yellow with greenish tinge at base. Back chestnut-brown and wings black with fine white barring. Breast pink and belly orangey-red, with vent a little lighter. Tail rufous, tipped with black, while undertail is white. Female drabber, with pink breast and red belly of male replaced by drab olive-brown. **DISTRIBUTION** Luzon, Bohol, Leyte, Samar, Mindanao. **HABITS AND HABITATS** Common inhabitant of primary and secondary forests from lowlands up to montane areas. Usually found in pairs in understorey, feeding on insects and fruits. Undulating flight when flushed. Has been found to nest in tree hollows. **SITES** Mt Makiling (Laguna), Rajah Sikatuna National Park (Bohol), PICOP (Surigao del Sur).

Male, Luzon

Spotted Wood Kingfisher ■ *Actenoides lindsayi* 25.4cm e

Male, Luzon

DESCRIPTION Medium-sized forest kingfisher with males and females differing in plumage. Male has dark green crown, short, pale rufous eye-stripe from lores to above eye, turquoise-blue eyebrow, black eye-stripe below eye extending to nape, forming collar, rufous moustache also extending to nape, forming another collar, turquoise malar stripe and pale rufous throat joining rufous moustache. Two-toned bill, with upper bill greenish-black and lower bill yellow-orange. Upperparts dark green with white spots; underparts buff-white with upper belly having dark green spots; green rump; greenish-brown tail. Female similar, with turquoise eyebrow and malar stripe of male replaced by green. Throat buff-white instead of rufous, and rufous collar replaced by black. **DISTRIBUTION** Luzon, Negros, Panay. **HABITS AND HABITATS** Common forest kingfisher found in lowland and mid-montane primary and secondary forests. Very active before dawn, giving out its explosive stuttering call. Perches quietly in understorey while searching for prey of insects and small vertebrates like lizards and skinks. **SITES** Subic Forest (Bataan), Mt Makiling (Laguna), Sierra Madre (Cagayan).

Hombron's Kingfisher ■ *Actenoides hombroni* 28cm e

Male

DESCRIPTION Medium-sized forest kingfisher with male and female differing in plumage. Male has dark blue crown and moustache, blackish-blue lores, rufous ear-coverts extending to nape forming rufous collar, white throat and bright red bill. Upperparts bluish-green with light rufous spots. Underparts from breast to vent rufous-brown. Rump light sky-blue and tail dark blue. Undertail pale rufous. Female very similar, but with blue cap and moustache replaced by dark green. **DISTRIBUTION** Mindanao. **HABITS AND HABITATS** Uncommon forest kingfisher found in primary and secondary forests, usually in middle- and high-level mountains of Mindanao, but does range down to lowlands. Perches motionless in dark understorey and typically calls pre-dawn. **SITES** Mt Kitanglad (Bukidnon), PICOP (Surigao del Sur).

Stork-billed Kingfisher ■ *Pelargopsis capensis* 33cm

DESCRIPTION Large kingfisher with huge bill. Head and underparts rufous. Big, stocky head with massive bright red bill. Back, wings and uppertail greenish-blue, rump lighter aqua-blue and legs orange-red. The race *gigantea* has an off-white head and upperparts. **DISTRIBUTION** For *gouldi* Palawan, Mindoro, Lubang; for *gigantea* Luzon, Mindanao, Samar, Leyte, Negros, Panay, Bohol, Tablas, Sibuyan, Bohol. **HABITS AND HABITATS** The largest of the Philippine kingfishers, found in coastal areas near mangroves and inland rivers. Uncommon and perches on exposed branches and rocks while looking for prey of crabs and fish. **SITES** Puerto Princesa Underground River Subterranean Park (Palawan), Sablayan (Mindoro).

Adult, Palawan

White-throated Kingfisher ■ *Halcyon smyrnensis* 26.5cm

DESCRIPTION Medium-sized kingfisher with dark chestnut head, back and underparts. Big, bright red bill and diagnostic small white throat. Upperparts and wings bright blue with black wing-coverts. Breast, belly and vent lighter chestnut-brown, and tail blue. In flight white patch on wings is clearly visible. **DISTRIBUTION** Throughout the Philippines except Palawan. **HABITS AND HABITATS** Commonly found in a wide variety of habitats, from streams and rivers in open country to forest edges. Favours exposed perches like telephone and power lines, and exposed branches while looking for prey. Varied diet, including fish, lizards, insects and amphibians. The race occurring in the Philippines is quite distinct from races in other countries, in having a very restricted white throat. **SITES** Subic Forest (Bataan).

Winchell's Kingfisher *Todiramphus winchelli* 24cm e

DESCRIPTION Medium-sized kingfisher, with male and female being slightly different. Male has blackish-blue crown adjacent to rufous lores connecting with blue stripe above eye, extending towards nape and forming collar. Sides of face black and nape rufous. Dark bluish-black wings and tail, and aqua-blue rump. Throat and rest of underparts white. Female similar, but has buff-rufous underparts. Black bill with off-white base to lower mandible. **DISTRIBUTION** Mindanao, Bohol, Samar, Leyte, Cebu, Tablas, Tawi-tawi. **HABITS AND HABITATS** Uncommon forest kingfisher that looks similar to more common Collared Kingfisher (see below), but varies greatly in its habits and habitats. Favours lowland primary and secondary forests, sometimes near forest streams. Active early in the morning, giving out its distinct call. Nests in abandoned termite mounds and feeds on lizards, insects and other invertebrates. **SITES** PICOP (Surigao del Sur); Rajah Sikatuna National Park (Bohol).

Female, Mindanao

Collared Kingfisher *Todiramphus chloris* 24cm

DESCRIPTION Medium-sized kingfisher with bright greenish-turquoise head, strong black bill, and white lores connecting with weak white supercilium. Black ear-coverts extend to nape, and white throat connects with broad white collar. Wings and tail turquoise-blue, with wing-tips black and rest of underparts white. Bill black with pearly-white base to lower mandible. **DISTRIBUTION** Throughout the Philippines. **HABITS AND HABITATS** The most common kingfisher in the Philippines, found in most areas but rarely in good forests. Habitats include mangrove areas, exposed reefs, mudflats and even wooded urban gardens. Very varied diet consisting of lizards, fish, crustaceans, snails, and insects such as mantises and grasshoppers. **SITES** LPPCHEA (Metro Manila), La Mesa Ecopark (Metro Manila), Balanga (Bataan).

Common Kingfisher

■ *Alcedo atthis* 15cm

DESCRIPTION Tiny kingfisher with dark blue-green head and nape adorned with small bright cobalt spots, rufous lores and ears, white patch behind ear, bluish-green malar stripe and white throat. Bill long and two toned, with black upper mandible and reddish-orange lower mandible. Bright aqua-blue back, rump and tail, and greenish-blue wings with small aqua spots. Upperparts rufous; legs orange. **DISTRIBUTION** Throughout the Philippines. **HABITS AND HABITATS** Common migrant kingfisher to the Philippines, favouring habitats associated with water. Usually seen perched on sticks and low-lying branches in coastal areas, mangroves, fish ponds and bigger rivers. Flies low over the water and gives out high-pitched call when disturbed. Often hovers over the water while hunting for fish. **SITES** LPPCHEA (Metro Manila), La Mesa Ecopark (Metro Manila), Candaba Marsh (Pampanga), Balanga (Bataan).

Philippine Dwarf Kingfisher ■ *Ceyx melanurus* 12–13cm ⓔ

DESCRIPTION Very small forest kingfisher with striking pink and rufous-lilac plumage. Rufous-orange head, orange lores and blue spot on top of white patch on neck behind ear. Purple-rufous upperparts with black wings tipped with bright blue spots. Throat white; bill bright red. Rest of underparts lighter rufous-pink, with vent becoming lighter. Bright red feet. **DISTRIBUTION** Luzon, Polillio, Mindanao, Samar, Leyte. **HABITS AND HABITATS** Uncommon forest kingfisher that is the Philippines' smallest. Found in primary and secondary forests and not usually associated with water. Despite bright plumage it is difficult to see, as it prefers to perch motionless near the ground in dark recesses in its forest home. **SITES** Sierra Madre (Cagayan), Polillio Island (Quezon), PICOP (Surigao del Sur), Pasonanca (Zamboanga).

Adult, Luzon

Indigo-banded Kingfisher ■ *Ceyx cyanopectus* 14cm e

DESCRIPTION Tiny kingfisher with male and female differing in plumage. Vibrant dark blue head with small cobalt-blue spots running down to nape. Rufous lores, deep brown eyes, and buff-and-white patch behind ear. Two-toned bill with black upper mandible and orangey-red lower bill. Scapulars dark blue, and wings darker blue with cobalt-blue tips in primaries. Back lighter cobalt-blue up to rump. Underparts from belly to vent orange-rufous. Legs and feet orangey-red. Male has two indigo-blue bands, with upper band running across breast and second broader band near belly; combined they form red heart shape on breast. Second indigo band continues on flanks. Female lacks upper indigo band. The race *nigrirostris*, found on Panay, Negros and Cebu, has all-black bill; male has single breast-band and sometimes second incomplete one. **DISTRIBUTION** Luzon, Mindoro, Negros, Panay, Cebu. **HABITS AND HABITATS** Uncommon kingfisher that prefers to perch on tops of exposed rocks and low-lying branches in clean, small freshwater streams in lowland forests. Dives head first into the water looking for its preferred prey of small crabs and fish. Nest is on suitable soil banks near forest streams. **SITES** Lagawe (Ifugao), La Mesa Ecopark (Metro Manila), Mt Makiling (Laguna), Villa Escudero (Quezon).

LEFT: *Male;* RIGHT: *Female*

Southern Silvery Kingfisher ■ *Ceyx argentatus* 15cm e

DESCRIPTION Small black-and-white kingfisher. All-black head with small, silvery-white spots running on sides of crown forming eyebrow, tiny white lores, white patch behind ear and all-black bill. Upperparts all black with wings displaying small, silvery-blue spots near scapulars. Back and rump very light cobalt-blue. Throat white and rest of underparts blackish cobalt-blue, with centre of belly all white. Bright red legs. **DISTRIBUTION** Mindanao. Closely related Northern Silvery Kingfisher *Ceyx flumenicola* on Samar, Leyte, Bohol. **HABITS AND HABITATS** Uncommon kingfisher that inhabits clean freshwater streams and pools in forests and forest edges, as well as small rivers near wooded areas. Prefers to perch on tops of rocks and branches near ponds, and dives head first into the water when hunting for small fish and crustaceans. **SITES** PICOP (Surigao del Sur).

Adult, Mindanao

Blue-tailed Bee-eater

■ *Merops philippinus* 29cm

DESCRIPTION Medium-sized bird with slim build. Olive-green crown, nape and upper back. Thin, bluish-white eye-stripe, black stripe across red eye and thin white patch under black stripe. Bill black and decurved. Buff chin adjacent to rusty-brown throat. Rest of upperparts greenish-blue, with primaries bright cobalt-blue tipped with black. Rump and tail sky-blue with long black central tail feathers. Underparts light greenish-olive with sky-blue vent and buff-white undertail. **DISTRIBUTION** Throughout the Philippines except Palawan. **HABITS AND HABITATS** Common bee-eater that favours open country, grassland, rice fields and marshy areas. Graceful flight, frequently hawking for insects on the wing, then landing on exposed branches, bushes or wires. Colonial nester, building several nest holes in soil banks. **SITES** Candaba Marsh (Pampanga), UP Los Baños (Laguna), fields throughout its range.

Blue-throated Bee-eater

■ *Merops viridis* 28cm

DESCRIPTION Medium-sized bee-eater with rufous-chestnut crown extending to upper back. Dark brown lores and stripe across red eye. Sides of face greenish-blue, lime-green throat and decurved black bill. Wings dark yellowish-green with darker tips, back and rump sky-blue, and tail vibrant sky-blue, including central tail feathers. Breast and belly area lighter lime-green, and vent has slight bluish tinge. **DISTRIBUTION** Throughout the Philippines except Palawan. **HABITS AND HABITATS** Common but range-restricted bee-eater of open areas close to forests and forest edges. Usually seen in small groups, sometimes perched on wires; flies out gracefully to catch various insects on the wing. Nests in colonies, with eggs laid in tunnel a metre or more into soil banks. **SITES** Subic Forest (Bataan).

Palawan Hornbill *Anthracoceros marchei* 61–71cm (e)

DESCRIPTION Medium-sized hornbill with almost entirely black plumage contrasting with large, creamy-white casque (raised hollow ridge characteristic of most hornbills). Lower bill has black base. Bare white skin around eye, and white gular patch adjacent to black base of lower bill. Upperparts have greenish-black gloss. Long, cream-white tail. Dark red eyes with long eyelashes, and dark grey legs. Immature birds have pinkish white casque. **DISTRIBUTION** Palawan. **HABITS AND HABITATS** Uncommon hornbill that is the only hornbill in the Palawan group, and is found in primary and secondary forests and forest edges. Quite vocal and can be seen travelling in family groups to feed on fruiting figs in the canopy. **SITES** Puerto Princesa Underground River Subterranean Park (Palawan).

Rufous Hornbill *Buceros hydrocorax* 89–94cm (e)

DESCRIPTION Large hornbill with impressive red bill. Dark-rufous crown and nape, black face including chin, and buff-white lower throat. Wings and back dark brown and upper breast dark rufous. Rest of underparts dark-brown, with lower belly and thighs light rufous. Long, creamy-white tail. Orange-red legs. Race *hydrocorax* has all-red bill, while *semigaleatus* and *mindanensis* have red bills with distal half off white. Upper mandible has huge casque. Immature birds have black bills, and buff-white head, neck and underparts. Upperparts and wings dark brown with buff-white fringes. **DISTRIBUTION** Luzon, Bohol, Samar, Leyte, Mindanao. **HABITS AND HABITATS** Uncommon hornbill that is the largest in the Philippines, favouring primary forests from coastal areas to mid-level mountains. Usually found in family groups, sometimes flying for large distances in search of fruiting trees. Like other Philippine hornbills, nests in tall trees, with male imprisoning female inside tree hollow, then cementing the opening to leave a hole just big enough for him to feed female and young. **SITES** Sierra Madre (Cagayan), PICOP (Surigao del Sur).

Adult, Luzon

Luzon Hornbill *Penelopides manillae* 45cm e

DESCRIPTION Small hornbill with male and female differing in plumage. Male has creamy-white head, and black face with buff-white facial skin around eyes. Chin buff-white; ear-coverts and throat black. Red eyes with long eyelashes. Bill black, with number of white-and-yellow notches depending on age; distal half of bill dark brown. Back and wings brownish-black with greenish gloss; rest of underparts off white with light yellow wash. Back and uppertail-coverts greenish-black; tail brownish-black with subterminal rufous band. Legs dark brown. Female lacks buff-white underparts, which are replaced by dark brown plumage. **DISTRIBUTION** Luzon, Marinduque, Catanduanes, Polilio. **HABITS AND HABITATS** The Philippines' most common hornbill, found in forests and forest edges in lowlands. Feeds almost exclusively on fruits, with a strong preference for figs, but occasionally eats small lizards and insects. Usually found in pairs or family groups, flying just below the canopy. Like other Philippine hornbills, a cavity nester. **SITES** Sierra Madre (Cagayan), Subic Forest (Bataan), Mt Makiling (Laguna).

Male

Writhed Hornbilll *Rhabdotorrhinus leucocephalus* 68–77cm e

DESCRIPTION Medium-sized hornbill, with male and female differing slightly in plumage. Male has dark rufous crown, and golden-buff face, neck and upper breast. Orangey-red skin around eye and throat, and dark brown eye-ring with long eyelashes surrounding red eye. Big red bill with small casque and dark grey, depressed grooves near lower bill-base. Back, wings and rest of underparts black with greenish gloss. Long tail buff with black terminal band. Female's upperparts and underparts black with greenish gloss; buff tail with black terminal band. **DISTRIBUTION** Mindanao, Dinagat, Camiguin Sur. **HABITS AND HABITATS** Locally common hornbill found in forests and forest edges. Usually occurs in small groups but sometimes seen in big flocks, especially when feeding on large fruiting trees. Prefers lowland forests and also nests in hollows of tall trees. **SITES** PICOP (Surigao del Sur), Baluno (Zamboanga), Pasonanca Natural Park (Zamboanga).

Male

Adult, Luzon

Coppersmith Barbet

■ *Psilopogon haemacephalus* 14.6cm

DESCRIPTION The only barbet occurring in the Philippines. Red crown, black face with yellow spots above and below the eye, strong black bill and yellow throat. Brown eyes surrounded by thin red skin. Nape bluish-green; rest of upperparts and wings olive-green with yellow edges. Red band below yellow throat, followed by thinner, brighter yellow band. Rest of underparts pale yellow with olive-green streaks. Red legs and feet. The *cebuensis*, *intermedia* and *homochroa* races differ by having red spots above and below eye and throat, instead of yellow spots. **DISTRIBUTION** Throughout the Philippines except Bohol and Palawan. **HABITS AND HABITATS** Favours lowland forests and forest edges. Usually perches high up in the canopy, often on exposed branches and dead snags. Found singly or in pairs, and sometimes in small groups when feeding on fruiting trees. Loud, far-carrying *poke-poke-poke*… call given for long periods. **SITES** Subic Forest (Bataan), Mt Makiling (Laguna), Tabunan (Cebu).

Philippine Pygmy Woodpecker ■ *Yungipicus maculatus* 14cm e

DESCRIPTION Small woodpecker with predominantly brown-black and white plumage. Dark crown and forehead; white stripe starting above eye and running down nape and into upper back. Brown lores and ear-patch forming brown stripe across eye, and alternating white-and-brown stripes from chin to throat. Upperparts blackish-brown with white streaks; rest of underparts buff-white with brown streaks. Male *validirostris* has red patch on nape, but most of the time this is not visible. **DISTRIBUTION** Throughout the Philippines except Palawan. **HABITS AND HABITATS** The Philippines' smallest woodpecker. Common in a wide variety of habitats from lowland to high-elevation forests, and even in wooded gardens of cities. **SITES** La Mesa Ecopark (Metro Manila), Subic Forest (Bataan), UP Los Baños (Laguna).

Female, Luzon

Spot-throated Flameback

Dinopium everetti 28–30cm e

DESCRIPTION Medium-sized woodpecker with male and female differing slightly in appearance. Male has bright red crown and short crest, creamy-white face with black stripe running across eye, black moustache and buff throat. Upperparts and wing-coverts vibrant olive-yellow, rump bright red and breast greenish-brown; rest of underparts buff with barring. Female similar but lacks crimson crown. Instead, crown is black with tips of crest tinged with red. **DISTRIBUTION** Palawan. **HABITS AND HABITATS** Common woodpecker found in forests and wooded open areas in lowland Palawan. Seen singly or sometimes in pairs. Previously lumped with Common Flameback complex of mainland Asia, but now elevated to full species status. **SITES** Sabang (Palawan), Coron (Busuanga).

Male

Buff-spotted Flameback

Chrysocolaptes lucidus 26–29cm e

DESCRIPTION Medium-sized woodpecker with golden-yellow upperparts. Male and female differ, with male having bright red forecrown and crest, and pale buff face and throat with undefined black streaks. Upperparts and wings golden-yellow; underparts paler and duller golden-buff with black speckling. Red eyes, dark grey bill and black tail. Female very similar to male, but has golden rather than red crown. **DISTRIBUTION** Mindanao, Bohol, Samar, Leyte, Basilan. **HABITS AND HABITATS** Uncommon woodpecker found in forests from lowlands up to high-elevation mountains. Usually occurs singly or in pairs. Quite vocal and will give call when foraging. Its drumming is loud and far carrying. **SITES** Rajah Sikatuna National Park (Bohol), Mt Kitanglad (Bukidnon).

Male

Male

Luzon Flameback

■ *Chrysocolaptes haematribon* 26–28cm e

DESCRIPTION Medium-sized woodpecker with dark crimson wings. Male has dark crimson crown, short crest, buff-white face with small black speckling, faint, thin moustachial stripe and broad black band running on sides of neck. Upperparts dark red; black tail. Underparts light greyish-brown, with breast heavily speckled with pale buff edged with black. Red eyes, greyish-black bill and fleshy-grey legs. Female has black crown and crest with small white spots. **DISTRIBUTION** Luzon, Polillio. **HABITS AND HABITATS** Common flameback found in suitable lowland forests and forest edges. Noisy and conspicuous woodpecker usually seen in pairs. Previously lumped within Greater Flameback complex. **SITES** Subic Forest (Bataan).

Male

Red-headed Flameback

■ *Chrysocolaptes erythrocephalus* 26cm e

DESCRIPTION Medium-sized woodpecker with bright red head and light-coloured bill. Male has completely red head with some black spots on ear-coverts, pinkish-red throat, dark eye with bluish eye-ring and pale yellowish bill. Greenish-gold back and wings, red lower back and rump, and chocolate-brown tail. Underparts buff-white with irregular black stripes and scales. Female similar to male, but head is mottled red and gold rather than solid red. **DISTRIBUTION** Palawan. **HABITS AND HABITATS** Uncommon flameback sharing same habitat as more common Spot-throated Flameback (see p. 87) in Palawan. Favours primary and secondary forests, and creeps up and down large tree trunks while looking for insects and grubs. **SITES** Sabang (Palawan). **CONSERVATION** Newly split woodpecker that has a very small population, believed to be declining due to rapid habitat loss.

Sooty Woodpecker

■ *Mulleripicus funebris* 30cm ⓔ

DESCRIPTION Medium-large woodpecker with generally all-dark plumage. Male has dark red forehead, face and malar stripe, and greyish-brown throat with small white spots. Rest of body flat black with bluish-black gloss. Female lacks dark red forehead, face and malar stripe. Pale yellow bill and yellowish eyes. **DISTRIBUTION** Luzon, Mindanao, Samar, Leyte. **HABITS AND HABITATS** Uncommon woodpecker favouring lowland primary and secondary forests and forest edges. Usually seen singly or in pairs. Feeds mainly on insects high in tall trees, but sometimes drops down closer to the ground. **SITES** Subic Forest (Bataan).

Male, Luzon

Great Slaty Woodpecker ■ *Mulleripicus pulverulentus* 48cm

DESCRIPTION Huge woodpecker with all-grey plumage. Male completely ashy-grey, with ashy-grey head, pinkish red malar stripe and dull yellow throat. Powerful, greyish-black bill, all-black tail and grey legs. Female lacks pinkish malar patch. **DISTRIBUTION** Palawan. **HABITS AND HABITATS** The largest woodpecker in the Philippines, and is uncommon. Favours forests and secondary growth and mangroves in lowlands, often creeping up and down exposed bare branches. Quite vocal, and forages for ants and other insects in tall trees, sometimes singly or in small family groups. **SITES** Sabang (Palawan).

Male (centre)

Wattled Broadbill ■ *Sarcophanops steerii* 16.5cm ⓔ

DESCRIPTION Medium-small broadbill with very distinctive and colourful plumage. Dark blackish-blue head accented by sky-blue wattle surrounding eye. Forecrown has tinge of purple, and white collar encircles neck. Light blue eyes and sky-blue bill with very broad base. Upper back and wings greyish-brown, with conspicuous white and bright yellow band. Lower back and tail brownish-rufous, with mauve rump and uppertail-coverts. Throat black and rest of underparts light mauve. Striking light blue legs and feet. Female differs from male in having all-white underparts. **DISTRIBUTION** Mindanao. **HABITS AND HABITATS** Rare in lowland primary and secondary forests, where it prefers to perch motionless in exposed branches. Sometimes joins mixed feeding flocks while foraging for insects. Melodious whistle call, and makes whirring noises when in flight and snapping sounds with bill when perched. **SITES** PICOP (Surigao del Sur), Pasonanca (Zamboanga).

LEFT: *Male*; RIGHT: *Female*

Male

Visayan Broadbill

■ *Sarcophanops samarensis* 16.5cm ⓔ

DESCRIPTION Previously lumped with Wattled Broadbill (see above) as a single species, but recently recognized as a separate species because of distinct morphological differences. Differs from Wattled Broadbill by presence of white-and-mauve patch in the wing instead of yellow. Also has grey collar rather than white, and mantle has a more purple tinge. Male has more mauve in underparts. Female has white vent. Bill, eye-wattle, legs and feet light blue. **DISTRIBUTION** Bohol, Leyte, Samar. **HABITS AND HABITATS** Rare broadbill that inhabits primary and secondary forests, frequenting the understorey and mid-levels of lowland forest. An insectivore, and sometimes joins mixed feeding flocks gleaning for small insects. **SITES** Rajah Sikatuna National Park (Bohol).

Whiskered Pitta

■ *Erythropitta kochi* 21.5cm ⓔ

DESCRIPTION Large pitta with diagnostic pinkish-white 'whiskers'. Crown orange-brown and rest of head darker brown. Brownish chin and throat, broad, bright cobalt-blue breast-band and distinctive scarlet belly. Back and flanks dull olive; wings mostly green with some blue and black. **DISTRIBUTION** Luzon. **HABITS AND HABITATS** Found in middle- to high-elevation forests; most commonly reported from the Cordillera and Sierra Madre mountains. The largest pitta in the Philippines – much larger than similar looking Red-bellied Pitta (see below). Forages on the ground and calls loudly from low perch. **SITES** Sierra Madre (Cagayan).

Philippine Pitta

■ *Erythropitta erythrogaster* 16.5cm

DESCRIPTION Small-medium pitta with all-brown head and throat. Upper breast black, lower breast cobalt-blue; may have indistinct white band at base of throat. Distinct scarlet-red belly and undertail-coverts. Upper back green transitioning to bright blue to uppertail; wings mostly blue with some black in primaries. Black bill and flesh-grey legs. **DISTRIBUTION** Throughout the Philippines. **HABITS AND HABITATS** Common but can be hard to spot. Forages on the ground in forests and forest edges, turning over leaves and scraping the soil. Gives long, mournful call, *waaaaaaupwoooooo*, especially in the morning while perched on small trees, or from tree stump near the ground. **SITES** Mt Makiling (Laguna), Puerto Princesa Underground National Park (Palawan), Pasonanca Park (Zamboanga), PICOP (Surigao del Sur).

Adult, Luzon

Adult, Palawan

Hooded Pitta

■ *Pitta sordida* 16.5cm

DESCRIPTION Small-medium pitta with all-black head and throat, and mostly green body and wings. Turquoise-blue wing-coverts and black primaries. Rump also turquoise-blue, and tail black tipped. Lower centre of belly black, and vent scarlet. Black bill. **DISTRIBUTION** Throughout the Philippines. **HABITS AND HABITATS** Found foraging in secondary growth and forest edges, this pitta is quiet for most of the year. More vocal during wet season, and its presence is signalled by loud *wup-wup* call made from low or mid-storey perch. When flushed may fly low for short distances, exposing obvious white wing-patches. **SITES** La Mesa Ecopark (Metro Manila), Sabang (Palawan), PICOP (Surigao del Sur).

Azure-breasted Pitta

■ *Pitta steerii* 20cm e

DESCRIPTION Medium-large pitta with black head, nape and tail, and emerald-green back. Wings green, azure and black. Chin and throat pure white. Underparts bright azure-blue with centre of belly jet-black, and lower belly and undertail-coverts bright red. Black bill. **DISTRIBUTION** Bohol, Leyte, Samar, Mindanao. **HABITS AND HABITATS** One of the most sought-after birds in the Philippines, this beautiful pitta is uncommon and often associated with limestone forests. Often seen foraging on the ground or singing from low perch. Call a loud *werp-werp-werp-werp-werp-werp*. **SITES** Rajah Sikatuna National Park (Bohol), PICOP (Surigao del Sur).

Golden-bellied Gerygone ■ *Gerygone sulphurea* 10cm

DESCRIPTION Tiny bird. Plain olive-brown above and pale yellow below except for undertail-coverts, which are whitish. Tail olive-brown with subterminal white spots. Bill short, straight and black; white lores and dark brown eyes. **DISTRIBUTION** Throughout the Philippines. **HABITS AND HABITATS** Commonly found singly or in small groups in wide range of habitats from forest edges, to open country, mangroves, parks, gardens, and commercial and residential areas. Very active in tree canopies, searching for insects. Loud, wheezy song more often heard than the bird is seen. **SITES** La Mesa Ecopark (Metro Manila), UP Los Baños Campus (Laguna), Olango (Cebu).

White-breasted Woodswallow ■ *Artamus leucorynchus* 19cm

DESCRIPTION Stocky, medium-sized bird. Head, wings and upperparts up to tail dark grey except for rump, which is white. Underparts all white. In flight wings are distinctly deltoid with white underwing. Strong, pale blue-grey bill and beady, dark brown eyes. **DISTRIBUTION** Throughout the Philippines. **HABITS AND HABITATS** Common, ranging from secondary forests to forest edges, open fields, towns and residential areas. Frequently seen perched on open branches, telephone and electric wires, and poles, singly or in tightly huddled groups, and flying out to catch prey on the wing. Very aggressive, often swooping down on humans approaching nests and mobbing other birds, even including larger birds of prey in flight. **SITES** Subic Forest (Bataan).

Female, Luzon

Bar-bellied Cuckooshrike

Coracina striata 30cm

DESCRIPTION Medium-large bird with all-grey plumage and distinctive black-and-white barring on underparts and rump. Tail darker grey. Light yellow eyes, and black bill and legs. Some subspecies may be all grey and lack barring pattern. Male and female usually differ in plumage. In *kochii* race found in Mindanao, male has all-grey throat and breast, with black-and-white barring on belly. Female has all-black and white barred underparts. In *striata* race found in Luzon, male is all-dark grey with black lores. Female is lighter grey with black-and-white barring on rump, belly and undertail-coverts. **DISTRIBUTION** Throughout the Philippines. **HABITS AND HABITATS** Uncommon, travelling through or over forests and forest-edge canopies, often in small groups. Noisy and very conspicuous, calling while perched or on the wing. **SITES** Subic Forest (Bataan), Sablayan (Mindoro), Puerto Princesa Underground River Subterranean Park (Palawan), Sabang (Palawan).

Blackish Cuckooshrike

Edolisoma coerulescens 25.5cm e

DESCRIPTION Medium-sized bird with entirely flat black plumage except for rump, which is slightly greyish. Black bill and legs, and brown eyes. **DISTRIBUTION** Luzon, Catanduanes and Marinduque. Believed to be extinct in Cebu. **HABITS AND HABITATS** Often found in small groups foraging or flying through the canopy. Occurs in lowland up to mid-montane forests. Very vocal, with several birds often calling at the same time. **SITES** Subic Forest (Bataan).

McGregor's Cuckooshrike ■ *Malindangia mcgregori* 21cm e

DESCRIPTION Small grey, black and white bird. Face, forehead, lores, throat and upper breast velvety-black. Crown, back and central tail feathers slaty-grey. Lower breast light grey, becoming white on belly up to vent. Outer-tail feathers black with white tips. Primaries black; wing has large, contrasting white wing-bar. Black bill, grey legs and brown eyes. **DISTRIBUTION** Mindanao. **HABITS AND HABITATS** Found in forests and forest edges at high elevations. Occurs singly or in small groups, sometimes with mixed flocks. **SITES** Mt Kitanglad (Bukidnon).

Pied Triller ■ *Lalage nigra* 16cm

DESCRIPTION Slim bird with black crown and upper back, greyish white underparts and supercilium, black eye-stripe, and grey lower back and rump. Tail black with broad white tips. Wings black with large white markings. Female more grey than male, with faint scaling pattern on breast. Black bill and legs. **DISTRIBUTION** Throughout the Philippines. **HABITS AND HABITATS** Common in forest edges, thickets, gardens and parks, singly, or in pairs or small groups. Often seen in the canopy, carefully inspecting foliage for insects. Not very conspicuous except for its distinct *che-che-che-che* call. **SITES** La Mesa Ecopark (Metro Manila), UP Los Baños Campus (Laguna).

LEFT: *Male*; RIGHT: *Female*

Green-backed Whistler ■ *Pachycephala albiventris* 16cm ⓔ

Adult, Northern Luzon

DESCRIPTION Medium-sized whistler with olive-green upperparts, wings and tail. Greyish breast, white belly, and yellow vent and undertail-coverts. Black bill and grey legs. **DISTRIBUTION** Luzon and Mindoro. **HABITS AND HABITATS** Found at all levels of forest, but most common at high elevations on Luzon. Often occurs in mixed feeding flocks. Inconspicuous and fairly quiet as it hunts for insects in the canopy. **SITES** Mt Polis (Mountain Province).

White-vented Whistler ■ *Pachycephala homeyari* 16.5cm

DESCRIPTION Plain-looking bird with uniformly reddish-brown upperparts and white underparts. Breast has light brown-grey wash. Black bill and grey legs. **DISTRIBUTION** Masbate, Negros, Panay, Sibuyan, Tablas, Mindanao, Sulu Archipelago. Near endemic, also found on small islands off Borneo. **HABITS AND HABITATS** Found singly or in pairs in lowland or mid-elevation forest. Fairly inconspicuous and its presence is usually detected by its call. Often perches upright momentarily before flying off. **SITES** Tabunan (Cebu), Twin Lakes (Negros).

Adult, Cebu

Yellow-bellied Whistler

Pachycephala philippinensis 15cm e

DESCRIPTION Medium-sized, slim bird with brownish-grey head and olive-green upperparts. Tail and wings olive-green with grey edges. Throat white transitioning into bright yellow breast and belly. Black bill and pink-grey legs. **DISTRIBUTION** Luzon, Samar, Leyte, Bohol, Mindanao. **HABITS AND HABITATS** Very inconspicuous except for loud whistling call. Moves about in undergrowth and canopy; sometimes forms part of mixed feeding flocks. **SITES** Mt Makiling (Laguna), Rajah Sikatuna National Park (Bohol).

Brown Shrike *Lanius cristatus* 19cm

DESCRIPTION Medium-sized bird with forehead and crown a varying shade of brownish-grey, and brown upperparts and tail. Lores and forehead brown with black 'bandit' mask through eye and faint white eyebrow. Tail and wings brown with some rufous markings. Throat white, gradually transitioning to orange on flanks and paler orange on belly. Female has faint barring on breast. Strong, hooked bill has darker upper mandible with black tip. **DISTRIBUTION** Throughout the Philippines. **HABITS AND HABITATS** Common and abundant migrant seen in all types of habitat, including beaches, mountains, forests, forest edges, parks, gardens and city centres. Aggressive and territorial, often chasing off other birds. Calls harshly throughout the day. Perches conspicuously in the open, where it hunts for insects and other small animals. Often impales prey on thorny branches or barbed wire. **SITES** La Mesa Ecopark (Metro Manila), Candaba (Pampanga), UP Los Baños (Laguna), Subic Forest (Bataan).

LEFT: *Immature*; RIGHT: *Adult*

Long-tailed Shrike

■ *Lanius schach* 24cm

DESCRIPTION Medium-sized shrike with long tail. All-black head, grey mantle, and orange back and rump. White throat, breast and belly, and orange flanks. Tail and wings mostly black, with white markings. Sharply hooked black bill. **DISTRIBUTION** Throughout the Philippines except Palawan. **HABITS AND HABITATS** Very aggressive; single birds and pairs are territorial over a small area. Often perches in the open on the lookout for insect and small animal prey. Calls out very harshly, but can also be heard singing melodiously for several minutes while perched. **SITES** Candaba Marsh (Pampanga), UP Los Baños (Laguna).

Adult, Luzon

Mountain Shrike

■ *Lanius validirostris* 20cm e

DESCRIPTION Shrike with uniform ashy-grey crown and upperparts. Wings and tail darker grey-brown. Forehead and lores black, extending into distinctive black mask across eyes. Underparts from throat to belly white; flanks pale orange. Black bill and blackish-grey legs. Easily confused with superficially similar Brown Shrike (see p. 97). **DISTRIBUTION** Luzon, Mindoro, Mindanao. **HABITS AND HABITATS** Uncommon at high elevations in mossy oak and pine forests, but also seen in forest edges and clearings, hunting from exposed perches. May overlap in habitat with other shrikes, but differentiated from them by uniform grey upperparts. **SITES** Mt Polis (Mountain Province), Mt Kitanglad (Bukidnon).

Ashy Drongo

■ *Dicrurus leucophaeus* 26cm

DESCRIPTION Dull bluish-grey bird, fairly uniform in colour except for face, wings and outer-tail feathers, which are darker. Black lores and forehead. Tail long and splayed. Black bill and legs, and reddish-orange eyes. **DISTRIBUTION** Palawan. **HABITS AND HABITATS** Noisy and conspicuous in forest edges, thickets and open areas on exposed perches. Often stays in specific area for some time, catching insects from favoured perches. May be found singly, or in pairs or small groups. Migrant races have been reported in Luzon. **SITES** Puerto Princesa Underground River Subterranean Park, Iwahig (Palawan).

Balicassiao

■ *Dicrurus balicassius* 27cm ⓔ

Adult, Luzon

DESCRIPTION All-black drongo displaying blue-green sheen in bright light. In the Visayas, the belly up to the vent is white. Short frontal crest, heavy and black bill, black legs and dull red eyes. **DISTRIBUTION** Luzon, Catanduanes, Marinduque, Mindoro, Cebu, Guimaras, Masbate, Negros, Panay. **HABITS AND HABITATS** Common in forests and forest edges. Very gregarious, often travelling in small, noisy groups. Has a variety of calls, and often mimics other species. Frequently remains perched while calling loudly. Sometimes associated with mixed feeding flocks. **SITES** Subic Forest (Bataan), Sierra Madre (Cagayan), Mt Makiling (Laguna), Tabunan (Cebu), Twin Lakes (Negros).

Adult, Luzon

Blue-headed Fantail

Rhipidura cyaniceps 18cm e

DESCRIPTION Medium-sized fantail with blue-grey head, chin, throat, breast and upper back. Black lores and blackish-blue ear-coverts. Rufous back, rump and outer-tail feathers, and black central tail feathers. Wings rufous except for primaries, which are black. Belly lighter orange. Black bill and dark brown eyes. **DISTRIBUTION** Luzon, Catanduanes. **HABITS AND HABITATS** Very conspicuous bird, often seen in mixed feeding flocks moving quickly through the forest understorey. Flies from one perch to another while flicking tail and calling cheerfully. Can also be seen in forest edges and secondary growth. Responds aggressively to imitations of its calls. **SITES** Mt Polis (Mountain Province), Mt Makiling (Laguna).

Philippine Pied Fantail *Rhipidura nigritorquis* 19cm e

DESCRIPTION Monochromatic-plumaged fantail with brownish-black head with distinctive thick white supercilium. Upperparts and wings dark brown-black. White chin, throat and underparts, with dark brown-black breast-band. Tail brown-black with feathers (except central tail feathers) tipped with broad white spot. Black bill and legs, and dark brown eyes. Recently split from **Pied Fantail** *R. javanica* of Southeast Asia including Borneo, Java and Sumatra. **DISTRIBUTION** Throughout the Philippines. **HABITS AND HABITATS** Very common fantail found in a variety of habitats, including parks, gardens, commercial and residential areas, mangroves, thickets and secondary growth. Extremely active, constantly moving from one perch to another while fanning tail. Very aggressive, often attacking pet cats and dogs. Has a very loud, melodious, metallic-sounding call. **SITES** La Mesa Ecopark (Metro Manila).

Black-and-cinnamon Fantail ■ *Rhipidura nigrocinnamomea* 15.8cm ⓔ

DESCRIPTION Medium-sized fantail with all-black head and broad white eyebrow. Uniformly cinnamon-coloured back, rump and wing-coverts; tail brighter red-orange. Primaries black with cinnamon edge. Belly and undertail-coverts cinnamon. In *nigrocinnamomea* race found in central and southern Mindanao (Mts Talomo and Apo) breast is white, but in *hutchinsoni* race found in western, eastern and northern Mindanao (Mt Kitanglad) breast is uniformly cinnamon. Black bill and dark grey legs. **DISTRIBUTION** Mindanao. **HABITS AND HABITATS** Common bird in high-elevation montane forests, including secondary growth and forest edges. Very active, with typical fantail behaviour of flitting from perch to perch in the canopy or understorey. Often a lead bird in mixed feeding flocks. **SITES** Mt Kitanglad (Bukidnon), Mt Talomo (Davao), Mt Apo (South Cotabato).

Short-crested Monarch

■ *Hypothemis helenae* 13.5cm ⓔ

DESCRIPTION Distinctive bird with vibrant cobalt-blue head, breast and back. Crown also bright blue, with short crest that is often raised. Lores, eye-ring, chin and throat form dark black mask. Wings deep blue-black; belly to undertail-coverts a stark contrasting white. Thin yet distinct cobalt-blue eye-ring, and beak that has black tip. Blue-grey legs. Female duller than male, with dark grey upperparts, wings and tail. **DISTRIBUTION** Luzon, Mindanao and Samar, including smaller islands like Camiguin Norte, Catanduanes, Polillo, Dinagat and Siargao. **HABITS AND HABITATS** Rare but can be locally common. Found foraging in the canopy often as part of a mixed feeding flock. Very active and vocal. **SITES** Camiguin Norte (Babuyan Islands), PICOP (Surigao del Sur).

Celestial Monarch

■ *Hypothemis coelestis* 15.4cm ⓔ

DESCRIPTION Very distinctive monarch. Male has bright cerulean-blue head and back. Crest of long, slender feathers a shade lighter than head, extending past nape; often held flat but sometimes raised when agitated or excited. Back, throat and breast bright azure-blue, becoming paler towards lower back and belly to undertail-coverts. Wings and tail lighter blue with dark edges. Thin, yellow-green eye-ring and brown eyes. Blue bill with dark tip, and blue-black legs. Female more drab and darker than male, with lighter belly. **DISTRIBUTION** Luzon, Samar, Dinagat, Mindanao, Basilan, Tawitawi, Sibuyan, Negros. **HABITS AND HABITATS** Rare in forests and forest edges, foraging in the canopy. Occurs singly or joins mixed feeding flocks with other monarchs and babblers. To birders, one of the most sought-after birds in the Philippines. **SITES** PICOP (Surigao del Sur).

Female, Luzon

Rufous Paradise Flycatcher

■ *Tersiphone cinnamomea* 21.5–30.5cm

DESCRIPTION Bright orange-red head, body and tail. Vivid sky-blue eye-ring, and blue-grey bill and legs. Gape light jade. Male may have long central tail feather. Female has lighter belly than male. Crest may or may not be raised. In the Mindanao and Samar race, rufous colour is not as vibrant. **DISTRIBUTION** Reported all over the Philippines except Palawan group of islands, Masbate, Bohol and Leyte. Near endemic; also found on Talaud Island (Indonesia). **HABITS AND HABITATS** Uncommon in understorey of forests and secondary growth, and often heard before it is seen. Occurs singly, in pairs or as part of mixed flocks, calling loudly. Readily responds to imitations of its calls, or phishing. **SITES** Mt Makiling (Laguna), Sablayan (Mindoro), PICOP (Surigao del Sur).

Citrine Canary-flycatcher

Culicicapa helianthea 14.6cm

DESCRIPTION Small flycatcher with yellowish-olive head and back, yellow rump, dark brownish-olive wings and tail with yellow edges. Throat and underparts bright yellow. Thin yellow eye-ring and lores. Tiny beak with dark horn upper mandible and dark brownish-orange lower mandible. Brownish-orange legs. **DISTRIBUTION** Throughout the Philippines except Mindoro, Marinduque, Masbate, Samar, Bohol and Basilan. **HABITS AND HABITATS** Usually seen in montane forests, but also reported in lowland forests. Occurs alone or in pairs. Perches upright, flying out to catch insects and returning to same perch before moving on. Can form parts of mixed flocks. **SITES** Mt Polis (Mountain Province), Mt Kanlaon (Negros).

Elegant Tit

Pardaliparus elegans 11.3cm e

DESCRIPTION Striking yellow, black and white bird. Black crown, throat, nape and mantle. Nape and mantle may have variable white or yellow spots. Bright yellow on cheek up to shoulders and underparts. Wings black with white spots forming distinctive wing-bars. Amount of yellow and white markings on wing and back may differ in the nine races. Short black tail with white outer feathers and tips. Black bill, grey legs and brown eyes. **DISTRIBUTION** Throughout the Philippines except in the Palawan group of islands, Marinduque, Bohol and Basilan. **HABITS AND HABITATS** Occurs in all forest types at all elevations. Usually found within mixed flock together with white-eyes, sunbirds, nuthatches and warblers. Very active and vocal, moving about in all levels of the canopy, searching the foliage and branches for insects. **SITES** Mt Polis (Mountain Province), Mt Makiling (Laguna), Tabunan (Cebu), Mt Kitanglad (Bukidnon).

Male, Luzon

Palawan Tit ■ *Pardaliparus amabilis* 11.4cm ⓔ

DESCRIPTION Similar to Elegant Tit (see p. 103), which is absent in Palawan, but with all-black head, neck and throat, and bright yellow underparts. Mantle and back light yellowish-grey. Short black tail with white outer feathers and tips. Wings black with distinct white wing-bars. Black bill, dark grey legs and brown eyes. **DISTRIBUTION** Palawan. **HABITS AND HABITATS** Uncommon but very distinctive and conspicuous. Found singly, in small groups or as part of mixed flocks moving through the forest canopy. Also seen at forest edges and in secondary growth. Call a series of clear, cheerful whistles. **SITES** Puerto Princesa Underground River Subterranean Park (Palawan), Sabang (Palawan), Iwahig (Palawan).

Male

White-fronted Tit

■ *Sittiparus semilarvatus* 13.3cm ⓔ

DESCRIPTION Drab-looking bird with all-black or dark brown plumage except for white forehead and lores. Black bill and legs, and brown eyes. May have small concealed white spot on neck. Luzon races *snowi* and *semilarvatus* differ from Mindanao race *nehrkorni* by absence of white wing-patch. **DISTRIBUTION** Luzon, Mindanao. **HABITS AND HABITATS** Occurs in lowland to mid-elevation forests. Also found in forest edges and secondary growth. Usually stays in canopy but can be seen perched in the open on dead branches, singing loudly, especially in the morning. Call a series of sharp, high-pitched, metallic notes. Generally rare but may be locally common in some areas. **SITES** Subic Forest (Bataan), Sierra Madre (Cagayan). **CONSERVATION** Considered Near Threatened because of rapid habitat loss that is contributing to decreasing population numbers. Poorly studied and only reported regularly at sites mentioned above.

Adult, Luzon

Horsefield's Bush Lark

■ *Mirafra javanica* 13.3cm

DESCRIPTION Small lark with mottled dark brown and grey upperparts. Can be distinguished from Oriental Skylark (see below) by stouter bill and more rufous colour, especially on wings. Throat, breast and belly pale with heavy dark streaks on breast. Distinct pale eyebrow. Bill greyish with lighter lower mandible; legs flesh coloured. May raise short crest, which is often not otherwise visible. **DISTRIBUTION** Throughout the Philippines except Palawan. **HABITS AND HABITATS** Uncommon. Prefers foraging in open fields such as those with dried stubble on harvested rice paddies, and surrounding dikes and dirt roads. In courtship display flies up several metres in the air while fluttering wings and singing loudly. **SITES** UP Los Baños (Laguna).

Oriental Skylark

■ *Alauda gulgula* 15.2cm

DESCRIPTION Buff-white upperparts and breast heavily streaked with dark brown; belly pale and uniform. Erectile crest short and streaked. Slender, two-toned bill, with pale horn upper mandible and lighter lower mandible. Brown eyes. Pale eye-stripe extends to crescent bordering ear-coverts. Tail plain brown except for outer pair of feathers, which are white. Wing feathers brown with paler edges; lack of rufous feathers helps distinguish it from Horsefield's Bush Lark (see above). Pink legs. **DISTRIBUTION** Throughout the Philippines except Palawan. **HABITS AND HABITATS** Found walking singly in open habitats such as grassland and dry fields. Often runs away when approached, or when flushed flies low over the ground for short distances. Known for energetic courtship display, during which individuals rocket up to 30m or more into the sky, all the while fluttering and singing loudly, and often hovering in place. **SITES** Candaba (Pamapanga), UP Los Baños (Laguna).

Ashy-fronted Bulbul

■ *Pycnonotus cinereifrons* 19cm ⓔ

DESCRIPTION Plain-looking, slightly built bulbul. Crown and forehead dull brown with grey fringes. Ear-coverts dark brown with conspicuous white shafts-streaks below eye. Upperparts all brown with olive tinge. Dirty-white chin, pale grey breast and lighter grey belly with yellowish tinge; yellowish vent. Flight feathers and tail olive with yellowish edges. Dark grey bill and pinkish-grey legs. Brown eyes. Formerly subspecies of **Olive-winged Bulbul** *P. plumosus*. **DISTRIBUTION** Palawan. **HABITS AND HABITATS** Common in forest edges and secondary growth. Gurgling, melodius call. **SITES** Puerto Princesa Underground River Subterranean Park (Palawan), Iwahig (Palawan).

Palawan Bulbul

■ *Alophoixus frater* 22.7cm ⓔ

DESCRIPTION Medium-sized, heavily built bulbul. Rusty-brown crest, tail and wings. Lores, cheeks and ear-coverts grey with white streaks. White chin, throat and breast grey with yellowish wash. Belly and undertail-coverts yellow. Back and rump olive. Flesh-coloured legs and bright brown eyes. Bill grey and thicker than bills of other bulbuls in same habitat. Formerly subspecies of **Grey-cheeked Bulbul** *A. bres*. **DISTRIBUTION** Palawan. **HABITS AND HABITATS** Found in forests and forest edges, foraging within the canopy or undergrowth in pairs or small groups. Small crest may not be conspicuous at all times. Largest bulbul found in Palawan. **SITES** Puerto Princesa Underground River Subterranean Park (Palawan), Iwahig (Palawan).

Sulphur-bellied Bulbul

■ *Iole palawanensis* 17cm ⓔ

DESCRIPTION Small bulbul with pale yellow underparts. Throat lighter yellow, becoming darker on breast. Light grey streaks visible on breast. Head brown with fine white streaks extending to cheeks. Upperparts brown. Wings and tail brown with olive edges. Grey bill, becoming lighter on lower part, yellow eyes and yellowish-grey legs. **DISTRIBUTION** Palawan. **HABITS AND HABITATS** Occurs in forests and forest edges. Usually seen alone or in small groups, feeding on fruiting trees. **SITES** Puerto Princesa Underground River Subterranean Park (Palawan), Iwahig (Palawan).

Philippine Bulbul

■ *Hypsipetes philippinus* 22.3cm ⓔ

DESCRIPTION Medium-sized bulbul. Adult birds have greyish-brown head with fine white streaks, extending to nape. Same streaking on ears. Rest of upperparts olive-brown, becoming lighter on wings and tail. Rufous throat and breast, finely streaked with white. Rest of underparts whitish bordered by olive flanks. Black bill, reddish-brown eyes, and brown or grey legs. **DISTRIBUTION** Throughout the Philippines except Palawan, Mindoro, Gigantes, Guimaras, Masbate, Negros and Panay. **HABITS AND HABITATS** Occurs in forests and forest edges, up to 2,000m. Seen alone or in groups, preferring to stay in the forest canopy and understorey. Actively feeds on fruiting trees and has a variety of calls. **SITES** Mt Makiling (Laguna), Subic Forest (Bataan), PICOP (Surigao del Sur).

Adults, Mindanao

Yellowish Bulbul

Hypsipetes everetti 24.1cm e

DESCRIPTION Large bulbul with distinct yellow wash all over body. Upperparts greenish-yellow, becoming darker on wings. Top of head and upper back lightly streaked. Throat and cheeks reddish-brown, connecting to yellowish-brown chest and belly. Upper bill grey, lower bill bluish. Reddish-brown eyes and grey legs. **DISTRIBUTION** Samar, Leyte, Mindanao. **HABITS AND HABITATS** Inhabits lowland forests and forest edges, below 1,000m. Can be seen in groups or joining mixed flocks. May be quite noisy as it flies through trees in forests. **SITES** PICOP (Surigao del Sur), Camiguin Sur.

Barn Swallow

Hirundo rustica 17.8cm

DESCRIPTION Medium-sized swallow. Reddish forehead; chin and throat bordered by mottled black breast-band. Upperparts dark blue, becoming black towards wings and tail. Tail black and deeply forked, with row of white spots towards end. Black bill, dark brown eyes and black legs. Similar to resident Pacific Swallow (see opposite), which lacks black breast-band and has shallow, forked tail. **DISTRIBUTION** Throughout the Philippines. **HABITS AND HABITATS** Found in variety of habitats, including open sites, rural and urban areas, and forests. Usually seen in flight, catching insects in the air, and perched on electric wires and exposed branches. Often seen in large groups. **SITES** Candaba Marsh (Pampanga), Mt Makiling (Laguna), Bislig (Surigao del Sur), Rajah Sikatuna National Park (Bohol).

Pacific Swallow

■ *Hirundo tahitica* 12.7cm

DESCRIPTION Resident swallow similar to Barn Swallow (see opposite). Reddish forehead and throat, extending towards upper breast. Lower breast and belly greyish-white. Upperparts dark blue. Wings black. Tail black with shallow fork with row of large white spots towards end. Black bill, dark brown eyes and black legs. **DISTRIBUTION** Throughout the Philippines. **HABITS AND HABITATS** Inhabits a variety of habitats including open sites, rural and urban areas, and coastal regions. Can be seen in groups, flying to catch prey, and perched on electric wires and bare branches. **SITES** Subic Forest (Bataan), Candaba Marsh (Pampanga), Mt Makiling (Laguna), Bislig (Surigao del Sur), in variety of habitats throughout the Philippines.

Striated Swallow ■ *Cecropis striolata* 19cm

DESCRIPTION Large swallow with distinct red rump. Dark blue upperparts, becoming browner on wings. Top of head dark blue. Face buff finely streaked with black, extending around neck, throat and rest of underparts. Red wash on face and nape. Deeply forked black tail, and dark brown bill, eyes and legs. **DISTRIBUTION** Luzon, Masbate, Mindoro, Panay, Negros, Cebu, Bohol, Palawan. **HABITS AND HABITATS** Seen in open areas, flying to catch insects. Can also be observed perched on electric wires and bare branches. **SITES** Banaue (Ifugao), Mt Makiling (Laguna), Rajah Sikatuna National Park (Bohol).

Adult, from below

Adult, from above

Mountain Tailorbird ■ *Phyllergates cuculatus* 11.5cm

DESCRIPTION Small 'tailorbird' with distinctive bright rufous forehead extending to top of head. Nape grey connecting to olive-yellow back and rump. Tail olive-brown, becoming darker towards end. Wings olive-brown. Black lores extending as black eye-stripe. Short white eyebrow and light grey cheeks. Throat and breast white, connecting to yellow belly and undertail-coverts. Dark brown bill, brownish-grey eyes and pinkish-flesh legs. **DISTRIBUTION** In the Philippines, can be found throughout the country except on the islands of Mindoro, Panay, Masbate, Negros, Cebu, Bohol, Leyte and Samar. **HABITS AND HABITATS** Inhabits higher elevation forests and forest edges, above 800m. Can be seen alone, in pairs or joining mixed flocks. Quite difficult to see, preferring to stay in dense growths, where it can be heard calling. **SITES** Mt Polis (Mountain Province).

Rufous-headed Tailorbird ■ *Phyllergates heterolaemus* 11.5cm ⓔ

DESCRIPTION Small 'tailorbird' with bright rufous head. Rufous colour extends to nape, throat and upper breast. Rest of underparts bright yellow. Back yellowish-brown, becoming lighter on rump. Tail brown, becoming darker at end. Upper bill black, lower bill flesh. Brown eyes and pinkish-flesh legs. **DISTRIBUTION** Mindanao. **HABITS AND HABITATS** Found in higher elevation forests and forest edges, above 800m. Can be heard calling from dense tangles and undergrowth and quite challenging to see. May be seen alone, in pairs or joining mixed flocks. **SITES** Mt Kitanglad (Bukidnon).

Female

Philippine Bush Warbler ■ *Horornis seebohmi* 12.2cm e

DESCRIPTION Medium-sized bush warbler with white eyebrow and black eye-stripe. Top of head rufous-brown, which extends to rest of upperparts. White throat and underparts whitish-grey, becoming olive-brown on flanks. **DISTRIBUTION** Luzon. **HABITS AND HABITATS** More easily heard than seen, preferring to stay in thick undergrowth. Inhabits forests, including pine forests, at elevations above 800m. Can be seen alone or in pairs. **SITES** Mt Polis (Mountain Province).

Arctic Warbler

■ *Phylloscopus borealis* 12cm

DESCRIPTION Small warbler that is mostly brown all over body. Uniformly olive-brown upperparts, distinct white eyebrow and dark brown eye-stripe. Wings olive edged with white. Faint wing-bar can be seen. Underparts greyish-white. Upper bill dark grey and lower bill flesh coloured. Brown eyes and dark grey legs. **DISTRIBUTION** Throughout the Philippines. **HABITS AND HABITATS** Found in a variety of habitats, including forests, forest edges and cultivated areas. Overlaps with similarly marked **Japanese** and **Kamchatka Leaf Warblers** *P. xanthodryas* and *examinandus*, from which it can only be safely distinguished by its calls. **SITES** La Mesa Ecopark (Quezon City), Subic Forest (Bataan), Mt Makiling (Laguna), Candaba (Pampanga).

Philippine Leaf Warbler *Phylloscopus olivaceus* 11–12cm ⓔ

DESCRIPTION Small warbler with distinct yellowish eyebrow and broken eye-ring. Greyish-green head extending to rest of upperparts, including wings and tail, and pale yellow wing-bar. Underparts whitish with faint yellow streaks connecting to bright yellow patch under tail. Upper bill dark grey and lower bill flesh. Dark brown eyes and grey legs. **DISTRIBUTION** Mindanao, Negros, Bohol, Leyte, Samar. **HABITS AND HABITATS** Inhabits low- and mid-elevation forests, below 1,500m. Can be seen alone or joining mixed flocks, feeding on insects. Where they overlap on the island of Negros, similar to Lemon-throated Leaf Warbler (see below), which can be distinguished by yellow throat. **SITES** Rajah Sikatuna National Park (Bohol), PICOP (Surigao del Sur).

Lemon-throated Leaf Warbler *Phylloscopus cebuensis* 11–12cm ⓔ

Adult, Negros

DESCRIPTION Small warbler with distinct yellow throat. Yellow colour on throat connects with whitish breast. Light streaks can be seen on breast and belly. Bright yellow undertail and olive-green upperparts. Wings have same olive-green colour, but edged with yellow. Lighter yellow eyebrow and broken eye-ring. Upper bill dark brown and lower bill pale yellow. Dark brown eyes and grey legs. **DISTRIBUTION** Luzon, Negros, Cebu. **HABITS AND HABITATS** Quite commonly seen in forests and forest edges. Inhabits lowland to mid-elevation forests up to 1,800m. Found alone or joining mixed flocks. **SITES** Sierra Madre (Cagayan), Mt Makiling (Laguna), Mt Kanlaon (Negros), Tabunan (Cebu), Alcoy (Cebu).

Negros Leaf Warbler

■ *Phylloscopus nigrorum* 10–11cm (e)

Adult, Luzon

DESCRIPTION Small warbler that is mostly olive-brown all over. Underparts pale yellow. Tail yellow edged with white. Head and upperparts olive-brown, including wings. Faint white eyebrow. Upper bill dark brown and lower bill flesh coloured. Chestnut-brown eyes and grey legs. **DISTRIBUTION** Luzon, Negros, Panay, Mindoro, Mindanao. **HABITS AND HABITATS** Inhabits higher elevation forests, above 800m. Often joins feeding flocks, actively feeding on insects among tree branches. **SITES** Mt Polis (Mountain Province), Mt Kitanglad (Bukidnon).

Clamorous Reed Warbler ■ *Acrocephalus stentoreus* 17.5cm

DESCRIPTION Large, chunky warbler that is mostly pale brown and with long, rounded tail. Upperparts light brown, becoming darker towards rump. Wings brown edged with black. Light brown eyebrow and short crest on top of head can be seen in some birds. Throat buff-white, becoming darker towards breast and belly. Brown bill, light brown eyes and grey legs. **DISTRIBUTION** Luzon, Bohol, Leyte, Mindanao. **HABITS AND HABITATS** Inhabits open country and grassland. Can be heard singing from hidden perch, but also sings out in the open, on blades of grass or tops of trees and bushes. May be seen alone or in pairs. **SITES** Candaba Marsh (Pampanga), Bislig Airfield (Surigao del Sur).

Long-tailed Bush Warbler ■ *Locustella caudata* 18cm ⓔ

DESCRIPTION Small-medium ground dweller with distinct long, bushy tail, made up of loose tail feathers. Top of head dark brown, with brown extending towards nape. Rest of upperparts dark reddish-brown. Faint grey eyebrow; grey face with black lores extending as faint eye-stripe. Throat and chest greyish-white, with sides mottled with black. Rest of underparts dark reddish-brown. Upper bill black and lower bill grey. Reddish-brown eyes and brown legs. **DISTRIBUTION** Luzon, Mindanao. **HABITS AND HABITATS** Inhabits forests and forest edges above 700m. Quite difficult to see as it scurries on the forest floor. Can be seen alone or in pairs. Most often detected by scratchy song. **SITES** Mt Polis (Mountain Province), Mt Kitanglad (Bukidnon), Mt Talomo (Davao).

Adult, Luzon

Striated Grassbird ■ *Megalurus palustris* 26.6cm

DESCRIPTION Large grassbird with long brown tail. Adult birds mostly light brown all over with streaks on top of head, nape and back. Streaks larger and heavier on back and wings. Faint white eyebrow, and white throat and breast. Streaks on chest seen to form band. Belly buff bordered with brown streaks. Upper bill brown and lower bill grey. Brown eyes and legs. **DISTRIBUTION** Throughout the Philippines. **HABITS AND HABITATS** Found in grassland, open country and rice fields; also in grassy patches in cultivated areas. Likes to perch on exposed branches, tall blades of grass and even electric wires. Can be seen flying upwards and gliding back down, calling noisily. **SITES** Candaba Marsh (Pampanga), rice fields throughout the Philippines.

Zitting Cisticola

Cisticola juncidis 11cm

DESCRIPTION Small bird with heavy dark brown streaks on top of lighter brown head. Upperparts light brown mottled with dark brown. Orange-brown rump connecting to dark brown tail with buff edges. White throat and belly bordered by buff flanks. Upper bill black and lower bill flesh. Brown eyes and flesh legs. **DISTRIBUTION** Throughout the Philippines. **HABITS AND HABITATS** Found in open country, grassland and rice fields. Can be seen perched on tall blades of grass or branches close to the ground. **SITES** Candaba Marsh (Pampanga), rice fields throughout the Philippines.

Golden-headed Cisticola

Cisticola exilis 10m

DESCRIPTION Small, with male in breeding plumage developing bright rufous colour on top of head. Back brown with dark brown streaks. Rump rufous and tail dark brown, becoming rufous at tip. Wings dark brown with greyish-brown tips. Throat buff and breast brown. Belly white. In non-breeding plumage difficult to separate from Zitting Cisticola (see above), except by call. **DISTRIBUTION** Throughout the Philippines. **HABITS AND HABITATS** Inhabits fields and grassy areas. Can be seen perched on tall blades of grass or tops of trees. **SITES** Candaba Marsh (Pampanga), IRRI (Laguna), Bislig Airfield (Surigao del Sur), rice fields throughout the Philippines.

Trilling Tailorbird ■ *Orthotomus chloronotus* 12.8cm ⓔ

DESCRIPTION Small tailorbird with all-rufous head. Same as **Philippine Tailorbird** *O. castaneiceps*, but has a green mantle. **DISTRIBUTION** Northern and central parts of Luzon. **HABITS AND HABITATS** Found in forests and forest edges. Can be heard calling from unseen perch in dense tangles. Quite difficult to see in the open. **SITES** Subic Forest (Bataan), Sierra Madre (Cagayan).

Rufous-fronted Tailorbird ■ *Orthotomus frontalis* 12.8cm ⓔ

DESCRIPTION Distinct rufous forehead, extending to cover eyes. Top of head grey; olive-yellow back. Tail rufous, becoming darker towards end. Throat and breast white with grey streaks. Belly white bordered by pale yellow flanks. **DISTRIBUTION** Bohol, Leyte, Mindanao, Samar, Basilan. **HABITS AND HABITATS** Inhabits forests and forest edges, preferring dense growth in forest understorey. Often heard calling, but difficult to see. **SITES** Rajah Sikatuna National Park (Bohol), PICOP (Surigao del Sur).

Grey-backed Tailorbird ■ *Orthotomus derbianus* 12.5cm e

DESCRIPTION Same as **Philippine Tailorbird** O. *castaneiceps*, but has distinct grey back. **DISTRIBUTION** Luzon. **HABITS AND HABITATS** Found in lowland forests and forest edges, preferring the understorey and dense tangles. Quite difficult to see as it flits deep in brush areas. Can be heard calling from unseen perch. **SITES** Mt Makiling (Laguna), La Mesa Ecopark (Quezon City).

Rufous-tailed Tailorbird ■ *Orthotomus sericeus* 12.8cm

DESCRIPTION The only tailorbird found on Palawan. Adult has rufous head, covering eyes and extending towards nape. Grey back connects to rufous tail and greyish-brown wings. Underparts greyish-white with light grey streaks on breast. Upper bill brown and lower bill buff. Brown eyes and flesh-coloured or light brown legs. **DISTRIBUTION** Palawan. **HABITS AND HABITATS** Found in forest edges, scrub areas, mangrove areas and even near shoreline. Calls loudly and frequently from low perch. Prefers tangles, but also forages out in the open. Can be seen alone, in pairs or in small groups. **SITES** Puerto Princesa Underground River Subterranean Park (Palawan), Iwahig (Palawan).

Black-headed Tailorbird ■ *Orthotomus nigriceps* 12cm e

Male

DESCRIPTION Small tailorbird with black head. Black colour extends to nape, throat and upper breast. Thick white eyebrow. Lores and eye-ring also white. Upperparts olive-green. Wings olive-green edged with black. Underparts grey. Female has white throat and belly mottled with grey. **DISTRIBUTION** Mindanao. **HABITS AND HABITATS** Inhabits forests below 1,000m. Quite difficult to see, preferring to stay in dense undergrowth. **SITES** PICOP (Surigao del Sur).

Pin-Striped Tit-Babbler ■ *Macronus gularis* 14cm

DESCRIPTION Small tit-babbler. Top of head orange-brown. Wings, rump and tail have same orange-brown colouration; grey back. Face grey, connecting to pale yellow throat that has thin black streaks. Breast and belly greyish-yellow bordered by grey flanks. Thin white stripe on bend of wing can be seen. Grey bill, orange eyes and greenish-grey legs. **DISTRIBUTION** Palawan. **HABITS AND HABITATS** Found in forests and forest edges, below 1,500m. Forages noisily and actively close to the ground, preferring dense areas. **SITES** Puerto Princesa Underground River Subterranean Park, Iwahig (Palawan).

Brown Tit-Babbler ■ *Macronus striaticeps* 14.6cm ⓔ

Adult, Mindanao

DESCRIPTION Medium-small babbler with black face streaked with white. Upperparts rufous-brown with long white streaks running down centre, extending towards tail. White throat; rest of underparts rufous with short white streaks. **DISTRIBUTION** Samar, Leyte, Bohol, Mindanao. **HABITS AND HABITATS** Shy bird but calls noisily while travelling in small groups close to the ground. **SITES** Rajah Sikatuna National Park (Bohol), PICOP (Surigao del Sur).

Striated Wren-Babbler ■ *Ptilocichla mindanensis* 16cm ⓔ

DESCRIPTION Medium-sized ground-babbler. Top of head dark brown with faint buff streaks, extending towards forehead. Upperparts reddish-brown with white streaks towards lower back. Short brown tail. Wings reddish-brown. White lores and eyebrow, and greyish-brown cheeks. Throat white with black streaks on both sides. Breast and belly have distinct black-and-white streaks, connecting with reddish-brown flanks, also streaked with white. Upper bill black and lower bill grey. Reddish-brown eyes and pale brown legs. **DISTRIBUTION** Mindanao, Bohol, Leyte, Samar. **HABITS AND HABITATS** Found in dense undergrowth of lowland forests. Quite difficult to see, and usually alone or in pairs. Generally stays close to the ground in noisy groups. **SITES** Rajah Sikatuna National Park (Bohol), PICOP (Surigao del Sur).

Falcated Wren-Babbler ■ *Ptilocichla falcata* 19.3cm e

DESCRIPTION Medium-sized ground-babbler with striking plumage. Top of head dark reddish-brown, with colour extending to nape. Long feathers on dark brown back, streaked with white. Wings and tail dark reddish-brown. Lores reddish-brown, connecting to eye-stripe extending towards nape; dark brown patch below eyes to cheeks. Thin black malar stripe can be seen on white chin. Chest and belly black with short white streaks, becoming heavier towards belly. Upper bill black and lower bill grey. Reddish-brown eyes and dark brown legs. **DISTRIBUTION** Palawan. **HABITS AND HABITATS** Found in dense areas of forest, usually perched close to the ground. Quite difficult to see as it hops and walks on the ground. **SITES** Puerto Princesa Underground River Subterranean Park (Palawan), Iwahig (Palawan).

Ashy-headed Babbler ■ *Malacocincla cinereiceps* 11.5cm e

DESCRIPTION Medium-sized babbler with distinct grey head extending to nape, and thin grey malar stripe. Rest of upperparts brown, including wings, getting darker towards rump and tail. Tail distinctly short, giving characteristic shape. Chin and throat white. Light brown breast-band and light brown flanks. Upper bill grey and lower bill pinkish. Reddish-brown eyes, and long and pinkish legs. **DISTRIBUTION** Palawan. **HABITS AND HABITATS** Found in lowland and mid-elevation forests and scrub areas, either alone in the dense undergrowth or hopping on the ground. **SITES** Puerto Princesa Underground River Subterranean Park (Palawan), Iwahig (Palawan).

Melodious Babbler ■ *Malacopteron palawanense* 19.8cm e

DESCRIPTION Medium-sized babbler with rufous forehead and grey face. Upperparts olive-brown and tail orange-brown. Throat white streaked with grey. Belly white and undertail orange-brown. Eyes pale yellow. **DISTRIBUTION** Palawan. **HABITS AND HABITATS** Shy bird that can be quite difficult to see. Can be heard calling from low trees and tangles. May be confused with similarly coloured Ashy-headed Babbler (see opposite), but prefers perching on trees instead of on the ground. **SITES** Puerto Princesa Underground River Subterranean Park (Palawan), Iwahig (Palawan).

Chestnut-faced Babbler ■ *Zosterornis whiteheadi* 14.5cm e

DESCRIPTION Medium-sized babbler with rich chestnut colour on face extending to forehead and chin, and broken white eye-ring. Top of head and nape grey, connecting to olive upperparts, including wings and tail. Underparts pale yellow. **DISTRIBUTION** Luzon. **HABITS AND HABITATS** Found in higher elevation forests. Actively forages noisily in large groups, and also joins mixed flocks. Noisy and skittish, hopping from branch to branch, and transferring as a flock from tree to tree. **SITES** Mt Polis (Mountain Province), Baguio (Mountain Province).

Luzon Striped Babbler ■ *Zosterornis striatus* 13.1cm ⓔ

DESCRIPTION Small babbler with heavily streaked underparts. Upperparts greenish-brown connecting to reddish-brown tail. Forehead dark brown to black. Face brown with thin, short black eye-stripe and black spot below eye; thin black malar stripe. White lores and eye-ring. Throat and breast white, connecting to yellowish belly with dark brown streaks. Legs pale green. **DISTRIBUTION** Luzon. **HABITS AND HABITATS** Found in the understorey of forests and forest edges. Forages actively, joining mixed flocks and forming small groups. **SITES** Sierra Madre (Cagayan).

Flame-templed Babbler

■ *Dasycrotapha speciosa* 13.2cm ⓔ

DESCRIPTION The most spectacular of the Philippines' endemic babblers, with distinct orange tufts on temples, just behind and above eye. Yellow forehead, lores and eye-ring. Top of head black; grey patches on ears. Nape pale yellow; yellow chin connecting to black throat. Rest of upperparts, including wings, olive-brown with pale white streaks. Tail also olive-brown. Breast bright yellow spotted with black. Belly and undertail olive-yellow. Yellow-orange bill, reddish-brown eyes and yellow-green legs. **DISTRIBUTION** Negros, Panay. **HABITS AND HABITATS** Inhabits lowland forests, preferring to stay in the understorey. Can be seen alone, in pairs or joining mixed flocks. **SITES** Mt Kanlaon (Negros), Twin Lakes (Negros). **CONSERVATION** Endangered Philippine endemic due to habitat loss within its very limited range.

Mindanao Pygmy Babbler

Dasycrotapha plateni 11.4cm ⓔ

DESCRIPTION Small babbler. Adult birds have dark brown face, including forehead and chin, with distinct white streaks. Reddish-brown nape connecting to lighter brown colour on back, with less distinct white streaks. Brown wings. Tail brown with paler edges. Throat and breast reddish-brown with white streaks. Belly and undertail white. Grey bill and legs. Eyes orange with whitish pupil. **DISTRIBUTION** Mindanao. **HABITS AND HABITATS** Found in canopy and understorey of forests and forest edges, where it actively forages for insects on tree branches and leaves. Often seen in small groups and joining mixed flocks. **SITES** PICOP (Surigao del Sur), Pasonanca Park (Zamboanga).

Visayan Pygmy Babbler

Dasycrotapha pygmaea 11.4cm ⓔ

DESCRIPTION Small babbler. Head and neck olive-brown with faint white streaks. Throat and upper breast grey with faint white streaks. Brown upperparts. Eyes orange with whitish pupil. **DISTRIBUTION** Samar, Leyte. **HABITS AND HABITATS** Inhabits understorey and canopies of forests and forest edges. Actively forages for insects, and can be seen in small groups and joining mixed flocks. **SITES** Samar Island Natural Park (Samar).

Golden-crowned Babbler ■ *Sterrhoptilus dennistouni* 14cm ⓔ

DESCRIPTION Small babbler with distinct golden-yellow crown. Yellow colour on top of head extends to forehead. Upperparts brown with white streaks. Brown wings. Tail brown with reddish-brown at tip. Brown face has distinct white streaks. Throat and rest of underparts yellowish-grey, becoming lighter towards belly. Black bill, reddish-brown eyes and grey legs with some yellow. **DISTRIBUTION** Northern part of Luzon. **HABITS AND HABITATS** Inhabits lowland forests and forest edges, below 1,000m. Can be observed actively and noisily foraging among the trees. Often seen in small groups and joining mixed flocks. **SITES** Sierra Madre (Cagayan).

Mindanao White-eye ■ *Lophozosterops goodfellowi* 13.3cm ⓔ

DESCRIPTION Medium-sized white-eye lacking distinct white eye-ring. Upperparts olive-green. Face darker than rest of body; greyish-throat connecting to yellowish-grey underparts. Wings and tail dark grey. Black bill, reddish-brown eyes and greenish legs. **DISTRIBUTION** Mindanao. **HABITS AND HABITATS** Inhabits montane mossy forests, above 1,250m. Can be seen in groups or joining mixed flocks. **SITES** Mt Kitanglad (Bukidnon).

Lowland White-eye ■ *Zosterops meyeni* 10.2cm

DESCRIPTION Classic white-eye with distinct white eye-ring, and predominantly yellow plumage contrasting with white belly and undertail. **DISTRIBUTION** Found on Luzon and Mindoro; outside the Philippines recorded only on Lanyu Island. **HABITS AND HABITATS** Prefers urban environments or more open forests in lowlands. Can be seen actively foraging on fruiting and flowering trees. **SITES** La Mesa Ecopark (Quezon City), UP Los Baños (Laguna).

Yellowish White-eye ■ *Zosterops nigrorum* 10.8cm ⓔ

DESCRIPTION Predominantly yellow white-eye with typical white eye-ring. Underparts all yellow, but lighter than upperparts. **DISTRIBUTION** Luzon, Catanduanes, Mindoro, Leyte, Negros, Panay. **HABITS AND HABITATS** Inhabits mid-elevation forests and forest edges, below 1,000m. Can be seen in groups or joining mixed flocks, flying in groups from tree to tree. May overlap with similar Lowland White-eye (see above); distinguishable by all-yellow underparts. **SITES** Sierra Madre (Cagayan), Mt Makiling (Laguna), Mt Kanlaon (Negros), Twin Lakes (Negros).

Adult, Negros

Warbling White-eye ■ *Zosterops japonicus* 10.8cm

DESCRIPTION Typical white-eye with olive-green upperparts, becoming lighter at rump. Yellow throat contrasting with white belly, which is bordered with yellow. **DISTRIBUTION** Found on most Philippine islands except Catanduanes, Samar, Masbate, Leyte, Cebu and Bohol. **HABITS AND HABITATS** Inhabits high-elevation forests and forest edges, including pine forests, above 1,000m. Can be found in groups or joining mixed flocks. **SITES** Mt Polis (Mountain Province), Mt Kitanglad (Bukidnon).

Philippine Fairy-bluebird ■ *Irena cyanogastra* 24.7cm e

Male, Luzon

DESCRIPTION One of two fairy bluebirds found in the Philippines. Predominantly black but with bright blue on top of head, extending to nape, and dark blue back connecting to bright blue rump. Tail bright blue, becoming darker towards tip. Throat and upper breast black, connecting to dark blue belly. Bill black and eyes bright red. **DISTRIBUTION** Luzon, Samar, Leyte, Bohol, Mindanao. **HABITS AND HABITATS** Inhabits lowland and mid-elevation forests, below 1,500m. Can be observed alone or in groups, calling noisily to each other in the forest canopy. **SITES** Subic Forest (Bataan), Sierra Madre (Cagayan), Mt Makiling (Laguna), PICOP (Surigao del Sur).

Sulphur-billed Nuthatch

Sitta oenochlamys 12.2cm ⓔ

DESCRIPTION The only nuthatch in the Philippines outside Palawan. Adult male has pale blue upperparts extending all the way to tail. Wings have same blue colour but edged with black. Forehead and lores black; thick black eyebrow extending to nape, and bare yellow skin surrounding each eye. Small white spot can be seen below lores. Female lacks black eyebrow of male. Yellow bill and eyes, and yellowish-green legs. **DISTRIBUTION** Throughout the Philippines except Palawan. **HABITS AND HABITATS** Inhabits forests, where it frequents the canopy and middle storey. Can be seen alone or joining mixed flocks, creeping up and down tree trunks and branches. **SITES** Mt Makiling (Laguna), Mt Kitanglad (Bukidnon), Mt Talomo (Davao).

Asian Glossy Starling *Aplonis panayensis* 20.2cm

DESCRIPTION Medium-sized starling with all-black body. Adult birds all-black with greenish gloss. Immature birds have white underparts streaked with black. Black bill and legs, and blood-red eyes. **DISTRIBUTION** Throughout the Philippines. **HABITS AND HABITATS** Found in a variety of habitats, including urban areas, forest edges and coconut plantations. Feeds on fruits and insects. Can be seen in small to large groups, calling noisily from trees, and electric wires and posts. **SITES** Rajah Sikatuna National Park (Bohol), Tabunan (Cebu), PICOP (Surigao del Sur), towns and cities throughout the Philippines.

LEFT: *Adult*; RIGHT: *Immature*

Apo Myna

■ *Basilornis mirandus* 33cm ⓔ

DESCRIPTION Large myna with distinct fine-feathered crest, and yellow skin surrounding eyes. Male and female similar, adults appearing mostly black with bluish-green gloss. Lower back and rump white. Wings black but lack bluish-green gloss of rest of body. Yellow bill and dark brown eyes. Legs can be black or greenish in some birds. **DISTRIBUTION** Mindanao. **HABITS AND HABITATS** Inhabits higher elevation forests above 1,250m. Can be seen alone, in pairs or in groups. Perches on exposed branches and tree trunks. **SITES** Mt Kitanglad (Bukidnon), Mt Talomo (Davao).

Coleto ■ *Sarcops calvus* 29.2cm

DESCRIPTION Large and distinctive species with spectacular bald, pinkish-flesh skin covering most of face. Male and female appear similar, with mostly black plumage contrasting with grey on nape, lower back and rump. Underparts also black, but with grey on flanks. Black bill and legs, and dark brown eyes. **DISTRIBUTION** Near-endemic species found throughout the Philippines except Palawan. Outside the Philippines found only on Bangi Island, Sabah, Malaysia. **HABITS AND HABITATS** Inhabits forests and forest edges, and can be seen feeding on fruiting and flowering trees, and also on insects. May be seen in pairs or joining mixed flocks. Call a distinct mechanical sound. **SITES** Subic Forest (Bataan), Mt Makiling (Laguna), Twin Lakes (Negros), Mt Kitanglad (Bukidnon), PICOP (Surigao del Sur).

Stripe-headed Rhabdornis ■ *Rhabdornis mystacalis* 15.8cm e

DESCRIPTION Striking species with brown-and-white stripes. Top of head dark brown streaked with white, extending towards nape and upper back. Rest of upperparts dark brown. Face dark brown extending towards neck; white eye-stripe. Underparts white with bold brown streaks on sides of breast. Females have lighter brown colouration than males. Black bill, dark brown eyes and shiny black legs. **DISTRIBUTION** Throughout the Philippines except Palawan, Mindoro and Cebu. **HABITS AND HABITATS** Inhabits forests and forest edges in lower elevations, below 1,200m. Can be seen actively foraging among tree branches, feeding on insects and fruiting trees. Also found in groups and joins mixed flocks. **SITES** Subic Forest (Bataan), Mt Makiling (Laguna), PICOP (Surigao del Sur).

Adult, Luzon

Stripe-breasted Rhabdornis ■ *Rhabdornis inornatus* 16.5cm e

DESCRIPTION Subdued plumage rhabdornis. Upperparts brown, becoming darker on top of head and tail. Face dark brown with thin white eyebrow and whitish-grey throat. Breast has distinct but diffuse streaking, extending to flanks and undertail. Belly white. Female has lighter brown colouring than male. Black bill, dark brown eyes and grey legs, sometimes with yellowish tint. **DISTRIBUTION** Negros, Panay, Samar, Leyte, Mindanao. **HABITS AND HABITATS** Inhabits mid-elevation forests above 800m, preferring the forest canopy. Can be seen actively feeding on insects and fruiting trees. Likes to perch on dead snags or palm stems, especially in the early morning. **SITES** Mt Kitanglad (Bukidnon), Mt Talomo (Davao).

Adult, Mindanao

Ashy Thrush

■ *Geokichla cinerea* 19.2cm ⓔ

DESCRIPTION Medium-sized thrush. Adult has slate-grey upperparts, including wings, which show two prominent white wing-bars. Tail grey with white tips. Eye-ring white and surrounds only half the eye; white lores and white moustache. Sides of throat white. Underparts white with black spots on chest extending and becoming less dense on lower breast and belly. Centre of belly white. Immature birds have buff wash on body and spots that are just starting to develop. Bill black, becoming lighter at base. Dark brown eyes and pearly-flesh legs. **DISTRIBUTION** Luzon, Mindoro. **HABITS AND HABITATS** Inhabits lowland to montane forests. Can be seen hopping on the ground, foraging for worms under leaves. When flushed, flies to perch close to the ground. **SITES** La Mesa Ecopark (Quezon City), Mt Makiling (Laguna).

Adult, Eastern Mindanao

Island Thrush

■ *Turdus poliocephalus* 22.6cm

DESCRIPTION Medium-sized thrush represented by several distinct races in the Philippines, which vary considerably in plumage. Generally, adult male has dark greyish-black body. Bill, eye-ring and legs orange. Some races have white belly mottled with rufous or brown (*mindorensis* and *malindangensis*). Female has lighter colouration than male. **DISTRIBUTION** Found on most Philippine islands except Palawan, Masbate, Cebu, Bohol, Leyte and Samar. **HABITS AND HABITATS** Inhabits montane forests, above 1,000m. Can be seen feeding on fruits and insects, preferring forest understorey. Usually seen alone. **SITES** Mt Polis (Mountain Province), Mt Kitanglad (Bukidnon).

Siberian Rubythroat

Calliope calliope 15.4cm

Breeding male

DESCRIPTION Adult male has distinct red throat lined thinly with black, white malar stripe and white eyebrow. Upperparts olive and wings brownish-olive with buff edges. Underparts white, bordered with brownish-grey, becoming lighter on flanks. Female lacks red throat, which is instead buff-white. White malar stripe and eyebrow less distinct in female than in male. Pale brown breast-band. Bill black, becoming lighter at base. Dark brown eyes and flesh-coloured legs. **DISTRIBUTION** Luzon, Mindoro, Masbate, Panay, Negros. **HABITS AND HABITATS** Inhabits grassland, open country and marsh areas. Hard to see as it likes to skulk in dense undergrowth, but can also be observed to come out into the open to forage on the ground. **SITES** Candaba Marsh (Pampanga).

Philippine Magpie-Robin

Copsychus mindanensis 19.7cm e

Male

DESCRIPTION Adult male has black upperparts and black head extending to throat and chest, with bluish gloss to black feathers. White belly clearly separated from black chest. Wings black with distinct broad white wing-patch, and long black tail. Female has similar markings to male, but grey rather than black colouration. Immature birds have mottled plumage. Black bill and legs, and dark brown eyes. **DISTRIBUTION** Throughout the Philippines except Palawan. **HABITS AND HABITATS** Prefers secondary growth forests, scrub and cultivated areas. **SITES** La Mesa Ecopark (Quezon City).

Male, Luzon

White-browed Shama

■ *Copsychus luzoniensis* 16–18 cm e

DESCRIPTION Adult male has black head, throat and upper breast, contrasting with long, thick white eyebrow extending to nape. Back is black connecting to rufous rump. Tail black with outer feathers tipped with white. Wings black with prominent white wing-bars. Rest of underparts white, bordered with rufous flanks. Female greyish-brown instead of black, and has grey throat. Immature birds similar to female, but have paler grey throat. The Visayan race *superciliaris* lacks white wing-bar and rufous rump. **DISTRIBUTION** Luzon, Masbate, Negros, Panay. **HABITS AND HABITATS** More often heard than seen as it skulks in the undergrowth of primary and secondary forests. Song a beautiful loud and fluid series of notes, often with mimicry. **SITES** Subic Forest (Bataan), Mt Makiling (Laguna), Mt Kanlaon (Negros).

White-vented Shama ■ *Copsychus niger* 18–21cm e

DESCRIPTION Both male and female almost entirely black, with distinct white vent extending down to undertail. Long tail. Immature birds lack bluish gloss and have brown around white belly. **DISTRIBUTION** Palawan. **HABITS AND HABITATS** Inhabits lowland forests, forest edges and also cultivated areas. Can be seen singing its loud song from low perch. May be observed alone or in pairs. **SITES** Puerto Princesa Underground River Subterranean Park (Palawan), Iwahig (Palawan).

Luzon Water Redstart

Phoenicurus bicolor 14.5cm e

DESCRIPTION Charismatic species on mountain rivers. Adult male has dark blue head, throat and breast. Same blue colour on back connects to chestnut rump and tail. Wings dark brown, almost black. Belly deep chestnut in colour. Lores black. Adult female has paler colouration than male, particularly in the chestnut colour. Immature birds grey with lighter grey underparts. **DISTRIBUTION** Luzon, Mindoro. **HABITS AND HABITATS** Found in streams and rivers at higher elevations, above 300m. Can be seen perched on rocks near streams and rivers, from where it forages for insects. **SITES** Mt Polis (Mountain Province).

Male

Pied Bush Chat

Saxicola caprata 12.6cm

DESCRIPTION Adult male mostly black with distinct white wing-bar, white rump and white undertail. Adult female brown, becoming darker towards rump and tail. Underparts lighter shade of brown. Immature birds have similar colouration to adult female as males develop the black plumage. Black legs and bill, and dark brown eyes. **DISTRIBUTION** Occurs on most islands except Samar and Palawan. **HABITS AND HABITATS** Inhabits grassy areas and open country. Perches in the open on stalks of grass or exposed branches, and often flicks tail while perched. Usually seen in pairs. **SITES** Candaba (Pampanga), Bislig Airfield (Surigao del Sur), grassland throughout its range.

Male

Blue Rock Thrush ■ *Monticola solitaries* 20.3cm

DESCRIPTION Medium-sized thrush. Adult male has slaty-blue upperparts, head, throat and breast, which contrast with maroon belly extending to undertail. Wings dark blue to black. Tail black. Female lacks blue and maroon colouration of male, and is instead blue-grey all over with scale pattern on underparts. Black bill and legs, and dark brown eyes. **DISTRIBUTION** Throughout the Philippines. **HABITS AND HABITATS** Found in a variety of habitats, including forests, streams, and cultivated and urban areas. Likes to perch on tops of rocks, roofs, building ledges and tree branches. **SITES** Subic Forest (Bataan), UP Diliman (Metro Manila).

LEFT: *Male*; RIGHT: *Female*

Rufous-tailed Jungle Flycatcher ■ *Cyornis ruficauda* 14.5cm

DESCRIPTION Medium-sized flycatcher. Adult has rufous-brown upperparts with more rufous rump and tail; dark brown wings edged with rufous. Lores grey and spot over ear greyish-brown. Underparts whitish-grey. Light brown breast-band can be seen. Immature birds have spots on upperparts. Black bill, brown eyes and grey legs. **DISTRIBUTION** Samar, Leyte, Bohol, Mindanao. **HABITS AND HABITATS** Found in understorey of forests and forest edges. Can be seen joining mixed flocks. Has been reported up to mid-elevation forests, not exceeding 1,000m. **SITES** Rajah Sikatuna National Park (Bohol), PICOP (Surigao del Sur).

Adult, Mindanao

Grey-streaked Flycatcher ■ *Muscicapa griseisticta* 12.8cm

DESCRIPTION Medium-sized flycatcher with distinctly streaked underparts. Usually hunts insects from prominent perch. Underparts with greyish-brown streaks on white breast and belly. Upperparts brownish-grey, and wings darker compared with rest of body. White lores with even white eye-ring. Black bill and legs, and brown eyes. **DISTRIBUTION** Throughout the Philippines. **HABITS AND HABITATS** Usually seen perched conspicuously on bare branches in forests, forest edges or cultivated areas. Sallies to catch insects in flight before returning to original perch or branch nearby. **SITES** Mt Makiling (Laguna), Subic Forest (Bataan), Mt Kitanglad (Bukidnon), Pasonanca Park (Zamboanga).

Asian Brown Flycatcher ■ *Muscicapa dauurica* 12.8cm

DESCRIPTION Medium-sized flycatcher that is mostly greyish-brown. Wings greyish-brown with white edges. Can be told apart from similar looking flycatchers by white throat without any streaks, and has faint greyish-brown breast-band. Rest of underparts white. Wing-tips extend beyond tail when perched. **DISTRIBUTION** Palawan, Samar, Mindanao. **HABITS AND HABITATS** Uncommon migrant. Likes to perch in the open in forests and forest edges. Like most flycatchers, sallies to catch insects in flight before returning to original perch or nearby branch, then flying out again to feed. **SITES** Puerto Princesa Underground River Subterranean Park (Palawan).

Little Slaty Flycatcher ■ *Ficedula basilanica* 12.5cm e

DESCRIPTION Small flycatcher that prefers dark forest understorey. Adult male has slate-blue head and upperparts, with distinct white spot on neck, behind eye. Underparts white. Faint grey streaks form breast-band. Female lacks blue colouration of male and has rufous upperparts, becoming lighter towards rump and tail. Underparts white with light brown breast-band. Black bill, brown eyes and light grey legs. **DISTRIBUTION** Leyte, Samar, Mindanao. **HABITS AND HABITATS** Skulks in the understorey of lowland forests. Perches near the ground. Calls very softly from hidden perch. **SITES** PICOP (Surigao del Sur), Pasonanca (Zamboanga).

Male, Mindanao

Palawan Flycatcher ■ *Ficedula platenae* 11cm e

DESCRIPTION Small flycatcher with distinct orange throat and tail. Adults have orange-brown colour on top of head extending to lower back. Wings brown, edged with orange. Throat and breast lighter orange. Belly white extending to undertail. **DISTRIBUTION** Palawan. **HABITS AND HABITATS** Can be seen in low- and mid-elevation forests, below 1,000m. Quite difficult to see in forest understorey, where it prefers thick bamboo. **SITES** Iwahig (Palawan).

Cryptic Flycatcher

Ficedula crypta 11.3cm

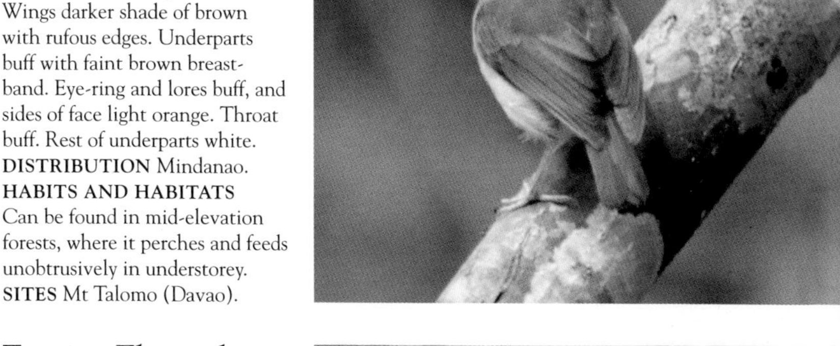

DESCRIPTION Small flycatcher with brown upperparts. Adult has brown head and nape connecting to more rufous rump and tail. Wings darker shade of brown with rufous edges. Underparts buff with faint brown breast-band. Eye-ring and lores buff, and sides of face light orange. Throat buff. Rest of underparts white. **DISTRIBUTION** Mindanao. **HABITS AND HABITATS** Can be found in mid-elevation forests, where it perches and feeds unobtrusively in understorey. **SITES** Mt Talomo (Davao).

Furtive Flycatcher

Ficedula disposita 11.3cm e

DESCRIPTION Small flycatcher with mostly brown body, but bright rufous base to tail. Adults have brown head and nape connecting to more reddish shade towards rump. Base of tail orange with brownish-black colour running down centre of tail. Tail ends brownish-black. Wings dark brown with lighter brown edges. Grey face with lighter grey lores and eye-ring. Throat and belly white, separated by greyish breast-band. **DISTRIBUTION** Luzon. **HABITS AND HABITATS** Found in forests, preferring dense growth and bamboo. Seen at elevations below 700m. Challenging to see as it sits motionless close to the ground. **SITES** Sierra Madre (Cagayan).

Little Pied Flycatcher ■ *Ficedula westermanni* 10.8cm

DESCRIPTION Small flycatcher with black-and-white plumage. Adult male has black upperparts with distinct white eyebrow, and white wing-patch. White edges of outer-tail feathers can be seen. Underparts white. Female lacks black-and-white colouring of male; instead upperparts are grey, becoming browner towards tail. Underparts greyish-white. Black bill and legs, and dark brown eyes. **DISTRIBUTION** Throughout the Philippines except Bohol, Cebu, Masbate, Panay and Samar. **HABITS AND HABITATS** Can be found in forests and forest edges, usually above 800m. **SITES** Mt Polis (Mountain Province), Mt Kitanglad (Bukidnon), Mt Talomo (Davao).

Male, Luzon

Turquoise Flycatcher ■ *Eumyias panayensis* 14.2cm

DESCRIPTION Medium-sized flycatcher with distinct pale blue colouration. Upperparts all blue. Black lores extend to surround eyes. Wings blue with brownish edges. Breast lighter blue colour, connecting to white belly. Black bill and legs, and dark brown eyes. **DISTRIBUTION** Throughout the Philippines except Palawan, Masbate, Samar, Cebu, Bohol and Panay. **HABITS AND HABITATS** Can be found in mossy forests, and mid-elevation forests and forest edges, above 800m. **SITES** Mt Polis (Mountain Province), Mt Kitanglad (Bukidnon), Mt Talomo (Davao).

LEFT: *Female, Mindanao*; RIGHT: *Male, Luzon*

Blue-breasted Blue Flycatcher ■ *Cyornis herioti* 15.1cm e

DESCRIPTION Medium-sized flycatcher with different shades of blue. Adult male has light blue forehead and eyebrow. Upperparts darker shade of blue. Throat and breast same dark blue shade. Chin and lores black. Sides of belly rufous. Centre of belly white. Wings blue, becoming darker towards tips. Female lacks blue colouration of male, instead having buff forehead and white eye-ring that is broken. Head and upperparts brownish-grey, connecting to rufous rump and tail. Wings brown, becoming lighter at edges, and rufous throat and breast. Belly whitish. Black bill, brown eyes and blackish-grey legs. **DISTRIBUTION** Luzon. **HABITS AND HABITATS** Quite uncommon and hard to see in lowland forests below 1,000m. **SITES** Sierra Madre (Cagayan).

Male, Northern Luzon

Palawan Blue Flycatcher ■ *Cyornis lemprieri* 16cm e

DESCRIPTION Medium-sized flycatcher. Adult male has dark blue upperparts. Black lores extend to cover face. White chin with small black spot at base of lower bill, and buff throat connecting to rufous chest. Centre of belly to vent white. Female lacks blue colouration of male and is mostly brown, becoming darker on wings and tail. Broken white eye-ring and white on chin. Chest rufous, connecting to white belly. Black bill, dark brown eyes and grey legs. **DISTRIBUTION** Palawan. **HABITS AND HABITATS** Inhabits primary and secondary forests, below 1,000m. Prefers to stay in forest understorey. Can be seen alone or in pairs. Overlaps with resident Mangrove Blue Flycatcher (see p. 140), with males being similarly marked. Females are easier to distinguish from each other, with female Palawan having uniform brown upperparts instead of blue, and female Mangrove Blue having similar blue colouration to male. **SITES** Iwahig (Palawan), Puerto Princesa Underground River Subterranean Park (Palawan).

Male

Mangrove Blue Flycatcher ■ *Cyornis rufigastra* 14.6cm

Male, Luzon

DESCRIPTION Medium-sized flycatcher. Adult male has dark blue upperparts, and black forehead, face and lores. Faded white eyebrow usually seen. Chin black, connecting to rufous throat and breast. Belly white, extending to undertail. Female distinguished from male by white lores, white chin and white spot below eye, and lacking black face. Black bill, brown eyes and grey legs. **DISTRIBUTION** Throughout the Philippines. **HABITS AND HABITATS** Found in lowland forests and forest edges. **SITES** La Mesa Ecopark (Quezon City), Tabunan (Cebu).

Olive-backed Flowerpecker ■ *Prionochilus olivaceus* 9.1cm e

DESCRIPTION Small but strikingly plumaged flowerpecker. Adult male has olive upperparts; wings darker olive with brown edges. Throat has distinct dark grey to black bands running down sides, extending down sides of chest and belly. White chin, centre of throat and centre of belly. Female has same markings as male, but has paler colouration and whitish-grey chest. Bill black with greyish lower bill. Bright red eyes and black legs. **DISTRIBUTION** Luzon, Samar, Leyte, Bohol, Mindanao. **HABITS AND HABITATS** Found in forests and forest edges, usually in forest understorey. Seen alone or joining mixed flocks, feeding on fruiting and flowering trees. **SITES** Sierra Madre (Cagayan), PICOP (Surigao del Sur).

Male, Mindanao

Palawan Flowerpecker ■ *Prionochilus plateni* 9cm ⓔ

DESCRIPTION Adult male has dark blue upperparts with distinct red patch on crown of head. Another red patch adorns breast. White malar stripes are striking, as is white chin connecting to yellow throat. Underparts yellow, becoming olive-green on flanks. Female lacks dark blue colouration and red patches of male. Olive upperparts. Black bill, becoming grey towards base. Red eyes and black legs. **DISTRIBUTION** Palawan. **HABITS AND HABITATS** Found in a variety of habitats, including cultivated and urban areas where there are suitable fruiting or flowering trees. **SITES** Iwahig (Palawan), Puerto Princesa Underground River Subterranean Park (Palawan), Coron (Busuanga).

Male

Olive-capped Flowerpecker ■ *Dicaeum nigrilore* 9.8cm ⓔ

DESCRIPTION Small flowerpecker with distinct long, curved bill and olive colour on crown of head. Same olive colour on rump and undertail. Black lores that extend to thick line below eye. Brown upperparts, becoming darker on wings. Throat and underparts whitish-grey. Black bill and legs. Red eyes. **DISTRIBUTION** Mindanao. **HABITS AND HABITATS** Found in mossy forests above 900m. Seen alone, but sometimes in larger numbers in suitable fruiting and flowering trees. **SITES** Mt Kitanglad (Bukidnon), Mt Talomo(Davao).

Flame-crowned Flowerpecker ■ *Dicaeum anthonyi* 9.6cm ⓔ

DESCRIPTION Small but chunky flowerpecker. The two races *anthonyi* and *kampalili* differ significantly. In Luzon race *anthonyi*, crown of head orange with yellow underparts. In *kampalili*, found on Mindanao, crown of head reddish-orange, belly greyish-white and undertail orange. Upperparts and wings bluish-black. Females have brown upperparts and olive underparts. **DISTRIBUTION** Luzon, Mindanao. **HABITS AND HABITATS** Found in forests and forest edges, preferring mossy forests above 800m. **SITES** Mt Polis (Mountain Province), Mt Kitanglad (Bukidnon).

LEFT: *Male*; RIGHT: *Female, Luzon*

Red-keeled Flowerpecker ■ *Dicaeum australe* 10cm ⓔ

DESCRIPTION Small flowerpecker with striking plumage combining bluish-black underparts and distinct red stripe running down centre of pale grey underparts. Female has same colouring as male but is paler. **DISTRIBUTION** Throughout the Philippines except Negros, Panay, Palawan and Mindoro, where it is replaced by **Black-belted Flowerpecker** *D. haematostictum*, with black border around red stripe on breast. **HABITS AND HABITATS** Usually seen in forest canopy, forest edges, open country, and some cultivated and urban areas. **SITES** Subic Forest (Bataan), Mt Makiling (Laguna), PICOP (Surigao del Sur).

Orange-bellied Flowerpecker ■ *Dicaeum trigonostigma* 9cm

DESCRIPTION Small flowerpecker with bright orange underparts. Male has yellowish orange throat finely streaked with red. Upperparts dark blue with a red spot on back. There are 11 races, which can be distinguished by the colour of the chin. *Xanthopygium* (Luzon), *pallidius* (Cebu) and *dorsale* (Masbate, Negros, Panay) have an all-yellow throat; *cinereigularis* (Bohol, Leyte, Samar, Mindanao), *isidroi* (Camiguin Sur) and *besti* (Siquijor) have yellow chin and grey throat; *sibuyanicum* (Sibuyan), *intermedium* (Romblon) and *cnecolaemum* (Tablas) have all-grey throat; *assimile* (Jolo, Tawi-Tawi) and *sibutuense* (Sibutu) have dark bluish-grey throat. Female has olive upperparts with yellow rump and yellow underparts. **DISTRIBUTION** Throughout the Philippines. **HABITS AND HABITATS** Seen in forests and cultivated areas below 1,500m. Seen feeding on fruiting and flowering trees. **SITES** Mt Makiling (Laguna), PICOP (Surigao del Sur).

Male, Negros

Buzzing Flowerpecker
■ *Dicaeum hypoleucum* 8.9cm ⓔ

DESCRIPTION Small flowerpecker with five races distributed across the country. The *cagayanensis* and *obscurum* races are characterized by olive upperparts and greyish underparts, while the *hypoleucum*, *mindanense* and *pontifex* races have brown upperparts and greyish-white underparts. Long, thin and slightly curved bill distinguishes it from similarly coloured flowerpeckers. **DISTRIBUTION** Throughout the Philippines except Palawan, Mindoro, Panay, Masbate, Negros and Cebu. *Cagayanensis* and *obscurum* found in Luzon, including Catanduanes, and *hypoleucum*, *mindanense* and *pontifex* found in Visayas and Mindanao regions. **HABITS AND HABITATS** Inhabits forests, scrub growth and cultivated areas. Seen in groups as well as joining mixed flocks, feeding on fruiting and flowering trees. Occurs at elevations below 1,500m. **SITES** Sierra Madre (Cagayan), Mt Makiling (Laguna), PICOP (Surigao del Sur), Mt Kitanglad (Bukidnon).

Adult, Mindanao

Pygmy Flowerpecker ■ *Dicaeum pygmaeum* 7.7cm ⓔ

DESCRIPTION Tiny flowerpecker. Adult male has dark head with greenish-blue gloss. Wings and tail have same dark colouration. Face dark grey, extending to nape. Upperparts dark grey, becoming lighter towards rump and tail. Throat and upper part of chest white, developing grey mottling towards belly. Undertail-coverts light yellow. White pectoral tufts can be seen. Female has paler colouration than male. Black bill, dark brown eyes and blackish-brown legs. **DISTRIBUTION** Throughout the Philippines except Panay. **HABITS AND HABITATS** Found in forests and forest edges, and secondary growth forests. Seen feeding noisily on fruiting and flowering trees, usually high up in the canopy. **SITES** Sierra Madre (Cagayan), Mt Makiling (Laguna), PICOP (Surigao del Sur), Mt Kitanglad (Bukidnon).

Adult, Palawan

Grey-hooded Sunbird ■ *Aethopyga primigenia* 10.8cm ⓔ

DESCRIPTION Small sunbird that has a distinct grey head and upper breast forming a hood. Black lores and some birds have a thin white line running down centre of throat. Rest of upperparts and wings olive green with a yellow rump. Belly is white bordered with yellow. Tail is grey with white tips. Bill is black. Eyes are bright red. Two races occur on Mindanao with the *primigenius* race having metallic green on forehead and ear-coverts. **DISTRIBUTION** Mindanao. **HABITS AND HABITATS** Usually seen in montane forests and forest edges between 1,000m and 1,700m. Seen alone, in pairs or joining mixed flocks feeding on the flowers of banana plants. In lower elevations is replaced by Metallic-winged Sunbird (see p. 147). **SITES** Mt Kitanglad (Bukidnon).

Purple-throated Sunbird

■ *Leptocoma sperata* 9.9cm

DESCRIPTION A number of races occur throughout the Philippines, each differing slightly in amount of red on underside and colour on upper back. Males differ significantly from females. Adult males have distinct metallic purple on throat and metallic green on top of head. Black lores extend towards ears. Upper back reddish; black in some races. Lower back metallic green connecting to brownish-black tail. Metallic green spot on shoulder can be seen. Rest of underparts bright red; yellow with red stripe in some races. Females lack colouration of males and have olive-brown upperparts and orange-brown wings. Underparts pale yellow. **DISTRIBUTION** Throughout the Philippines. **HABITS AND HABITATS** Found in a variety of habitats, including lowland forests, cultivated areas, urban areas and mangrove forests. Seen in pairs, and sometimes in small groups. Favours coconut plantations, where it feeds on coconut flowers. **SITES** Mt Makiling (Laguna), Rajah Sikatuna National Park (Bohol), PICOP (Surigao del Sur).

Male, Luzon

Copper-throated Sunbird

■ *Leptocoma calcostetha* 12.2cm

DESCRIPTION Medium-sized sunbird that typically inhabits mangroves. Adult male has metallic green on top of head, reaching just above eye. Face black, extending towards upper back. Green shoulder, rump and uppertail-coverts. Throat metallic copper colour framed by metallic purple stripe on both sides of throat. Purple colour extends down to breast. Pectoral tufts yellow but usually hidden. Belly and rest of underparts black. Black wings. Female lacks metallic colouring of male and is mostly olive on upperparts and yellow on underparts. Female has white eyebrow, faintly seen just above eye. Bill curved and black. Brown eyes and black legs. **DISTRIBUTION** Palawan. **HABITS AND HABITATS** Commonly seen in mangrove areas, but has also been reported in cultivated areas not far from shoreline. **SITES** Puerto Princesa Underground River Subterranean Park (Palawan), Rasa Island (Palawan).

Male

Olive-backed Sunbird ■ *Cinnyris jugularis* 11.4cm

Male, Palawan

DESCRIPTION The most widespread sunbird in the Philippines. Male has distinct metallic purple throat, extending to upper breast. Rest of underparts olive-green. Wings brownish with olive-green edges. Tail black. Bill curved and black. Eyes dark brown. Female lacks metallic purple throat. **DISTRIBUTION** Throughout the Philippines. **HABITS AND HABITATS** Common sunbird found in a variety of habitats, from cultivated areas to mangrove areas and forests. Can be seen feeding on flowering plants. Calls with a loud *che-wit* sound, or in faster and continuous, one-syllable *swit-swit-swit*. **SITES** Mt Makiling (Laguna), La Mesa Ecopark (Quezon City), Puerto Princesa Underground River Subterranean Park (Palawan), parks and gardens throughout the Philippines.

Flaming Sunbird ■ *Aethopyga flagrans* 9.5cm e

Male

DESCRIPTION Adult male has metallic green forehead, connecting with olive-green head and upperparts. Uppertail metallic green ending in black tail with hints of metallic blue. Chin metallic purple and throat black, extending to upper breast. Bright red patch visible in middle of black throat. Rest of underparts yellow with patch of orange in centre. Female lacks metallic colouring of male, and instead has olive upperparts, whitish chin, and greyish-white underparts with tinge of yellow on belly. **DISTRIBUTION** Luzon. **HABITS AND HABITATS** Inhabits forests and forest edges, up to 1,350m. Seen alone or joining mixed flocks. **SITES** Mt Makiling (Laguna).

Maroon-naped Sunbird ■ *Aethopyga guimarasensis* 9.5cm ⓔ

DESCRIPTION Adult male has red-orange wash on nape and upper back. Top of head metallic green. Lower back and wings orange-brown, becoming lighter on rump. Small green patch on rump connects with short black tail. Chin metallic purple. Black lores extend towards ear and throat. Centre of upper breast has bright orange patch. Belly and flanks yellow. Female lacks metallic colouration of male, and has olive-yellow upperparts, becoming lighter on wings. Underparts yellowish-grey, with light yellow belly. **DISTRIBUTION** Panay, Negros. **HABITS AND HABITATS** Inhabits forests and forest edges, up to 1,350m. Seen alone, in pairs or joining mixed flocks. **SITES** Twin Lakes (Negros).

Male

Metallic-winged Sunbird ■ *Aethopyga pulcherrima* 9.6cm ⓔ

DESCRIPTION Medium-sized sunbird with short tail. Male differs significantly from female. Adult male has metallic purple patch on forehead. Rest of upperparts olive-green, becoming lighter on rump. Metallic blue spot on ears. Wings olive-green; patch of metallic green near shoulder. Tail short and also metallic green. Throat and chest pale yellow, connecting to white belly. Female lacks metallic colouration of male, except on tail feathers. Black bill, bright red eyes and grey legs. **DISTRIBUTION** Leyte, Samar, Mindanao. Similar **Luzon Sunbird** *A. jefferyi* found in Luzon, and **Bohol Sunbird** *A. decorosa* found in Bohol. **HABITS AND HABITATS** Inhabits lowland to mid-elevation forests and forest edges, up to 1,500m. Seen alone or joining mixed flocks. **SITES** PICOP (Surigao del Sur).

Male

Lovely Sunbird ■ *Aethopyga shelleyi* 8–11cm e

DESCRIPTION Small sunbird with iridescent blue-green patch on crown of head. Head maroon, extending to nape, chin, sides of throat and upper back. Throat and chest yellow with heavy red streaks. Rest of underparts pale yellow. Wings olive. Yellow rump connects to metallic green tail. Bill curved and black. Black legs and dark brown eyes. Female lacks red colouring of male and is mostly olive. Throat and chest of female greyish-yellow. Belly yellowish-white. **DISTRIBUTION** Palawan. **HABITS AND HABITATS** Found in forests and cultivated areas. Usually seen feeding on fruiting trees and flowering plants. Aside from nectar, feeds on small insects. **SITES** Iwahig (Palawan), Puerto Princesa Underground River Subterranean Park (Palawan), Coron (Busuanga).

Male

Handsome Sunbird ■ *Aethopyga bella* 8–9cm e

DESCRIPTION Formerly lumped with Lovely Sunbird (see above), which is very similar but does not overlap in range. Adult male has metallic green on top of head. Cheek red, extending around nape and upper back. Wings olive-yellow. Yellow rump connects to green tail. Yellow throat and chest have fine red streaks forming band. Belly and undertail white. Female lacks colouration of male, and instead has olive-yellow upperparts. Underparts greyish-yellow, becoming lighter towards belly. Differs from Lovely Sunbird of Palawan by lack of red streaks on yellow breast. **DISTRIBUTION** Throughout the Philippines except Palawan. **HABITS AND HABITATS** Found in a variety of habitats, including forests, forest edges and cultivated areas, below 2,000m. Feeds on fruiting and flowering trees, as well as on insects. **SITES** Mt Makiling (Laguna), PICOP (Surigao del Sur).

Male

Magnificent Sunbird ■ *Aethopyga magnifica* 12.7cm ⓔ

DESCRIPTION Striking bright red sunbird. Adult male has red head, mantle and breast, with purple malar stripes running down sides of throat. Wings and belly black; yellow rump connects with blue-violet tail. Female lacks red colouring of male and is mostly olive. Wings and tail dark brown. Black bill, dark brown eyes and reddish-black legs. **DISTRIBUTION** Cebu, Negros, Panay, Sibuyan, Tablas. **HABITS AND HABITATS** Easiest to see feeding on fruiting trees and flowering plants. Seen in forests and also in cultivated areas. **SITES** Tabunan (Cebu), Alcoy (Cebu), Twin Lakes (Negros).

Male

Pale Spiderhunter ■ *Arachnothera dilutior* 15.2cm ⓔ

DESCRIPTION The only spiderhunter in Palawan. Upperparts entirely dark olive; paler grey on underparts. Distinct pale eye-ring. Bill long and curved, and black with grey base. **DISTRIBUTION** Palawan. **HABITS AND HABITATS** Found in forests, forest edges, mangroves and cultivated areas. Feeds on flowering plants such as banana plants and other ornamental species. **SITES** Puerto Princesa Underground River Subterranean Park (Palawan), Iwahig (Palawan).

Naked-faced Spiderhunter ■ *Arachnothera clarae* 17.2cm e

Adult, Luzon

DESCRIPTION Distinct pinkish skin below eye and grey lores. Upperparts olive-green. Wings olive-green with yellow edges. Throat and breast grey mottled with white, becoming paler towards belly. Bill long and curved. **DISTRIBUTION** Luzon, Samar, Leyte, Mindanao. **HABITS AND HABITATS** Found in forests and forest edges, as well as in areas with flowering trees and banana plants. Calls noisily as it feeds actively on flowers and blooms. **SITES** Mt Makiling (Laguna).

Cinnamon Ibon ■ *Hypocryptadius cinnamomeus* 15cm e

DESCRIPTION Common endemic found in higher elevation forests of Mindanao. Almost entirely cinnamon coloured, with wing feathers and outer-tail feathers having blackish edges. Underparts lighter cinnamon coloured, becoming grey on belly. Bill grey with darker shade on top mandible. Brown eyes and grey legs. **DISTRIBUTION** Mindanao. **HABITS AND HABITATS** Inhabits montane forests, above 1,000m. Usually seen joining mixed flocks, feeding on insects in the canopy, but also gleaning them from branches and trunks in the manner of a nuthatch. **SITES** Mt Kitanglad (Bukidnon), Mt Talomo (Davao).

Eurasian Tree Sparrow ■ *Passer montanus* 13.2cm

DESCRIPTION One of the most familiar birds in the Philippines. Distinctive head pattern combines black throat-patch running down to breast, chestnut cap and white cheeks with black spot. Rest of upperparts rufous and black. Underparts grey. **DISTRIBUTION** Throughout the Philippines. **HABITS AND HABITATS** Associated with human communities. Thrives on crumbs, leftover food and refuse that can be scavenged in towns, cities and other inhabited areas. **SITES** Found throughout the Philippines in places associated with human communities.

Green-faced Parrotfinch ■ *Erythrura viridifacies* 12.2cm ⓔ

DESCRIPTION Distinctive finch with all-green body that contrasts with distinct long, pointed red tail. **DISTRIBUTION** Highly nomadic Philippine endemic that irrupts in association with flowering bamboo. Most records from Luzon, but also recorded from Negros and Mindoro. **HABITS AND HABITATS** Rarely seen, but drawn to sites with flowering bamboo. Stays quite a long time in an area to feed on grain-like bamboo seeds while supply lasts, then disappears from site. **SITES** Sierra Madre (Cagayan). **CONSERVATION** Listed as Vulnerable, falling victim to the illegal pet trade.

Red-eared Parrotfinch ■ *Erythrura coloria* 10.4cm e

DESCRIPTION Small and dumpy bird. Green body, and striking blue-and-red face pattern. Tail bright red. **DISTRIBUTION** Mindanao. **HABITS AND HABITATS** Seen alone or in small groups, foraging close to or on the ground in forests and forest edges. Also seen in grassy areas and fields of wild sunflowers. **SITES** Mt Kitanglad (Bukidnon).

Scaly-breasted Munia ■ *Lonchura punctulata* 10.8cm

DESCRIPTION Small bird with dark brown face and head. Chest and upperparts white with scaly markings, extending to undertail. Bill grey, short and pointed. Eyes dark brown. Legs grey and tail short and brown. **DISTRIBUTION** Throughout the Philippines. **HABITS AND HABITATS** Commonly seen in grassland and rice fields, and other areas with tall grasses. Perches on blades of grass and leafless shrubs. Usually seen in flocks. **SITES** Candaba Marsh (Pampanga), Mt Makiling (Laguna), Bislig Airfield (Surigao del Sur), grassland throughout the Philippines.

Chestnut Munia

Lonchura atricapilla 10.9cm

DESCRIPTION Characteristic bird of agricultural areas, with black head and deep chestnut-coloured body. Belly black, extending to undertail. Bill grey, short and pointed. **DISTRIBUTION** Throughout the Philippines. **HABITS AND HABITATS** The National Bird of the Philippines before it was replaced by the rather more spectacular Philippine Eagle. Usually found in large flocks in rice fields, grassland and open country. Perches on blades of grass and rice stalks. **SITES** Candaba Marsh (Pampanga), Bislig Airfield (Surigao del Sur), Mt Kitanglad (Bukidnon), rice fields throughout the Philippines.

Eastern Yellow Wagtail

Motacilla tschutschensis 16.6cm

DESCRIPTION Non-breeding adult has greyish-brown head. Upperparts uniform olive, extending to brownish-black tail. Wings brownish-black with two wing-bars. Underparts pale yellow. White eyebrow. In breeding plumage underparts all yellow. Black bill, dark brown eyes and grey legs. Stockier build compared with the Grey Wagtail's (see p. 154). **DISTRIBUTION** Throughout the Philippines. **HABITS AND HABITATS** Commonly seen in groups in open country and rice fields, at all elevations. **SITES** Candaba (Pampanga), Mt Makiling (Laguna), Iwahig (Palawan), Mt Kitanglad (Bukidnon).

Immature

Adult

Grey Wagtail ■ *Motacilla cinerea* 17.6cm

DESCRIPTION Non-breeding adult has grey head and back. Rump yellow, extending to uppertail-coverts. Rest of tail black with light yellow edges. Wings dark brown. White eye-stripe and throat. Underparts yellow. In breeding plumage, develops solid black throat. Black bill, dark brown eyes and light brown legs. **DISTRIBUTION** Throughout the Philippines. **HABITS AND HABITATS** Seen along streams and areas associated with water. Walks continuously, habitually bobbing tail. Calls with loud, metallic note when flushed or in flight. **SITES** Subic Forest (Bataan), Mt Makiling (Laguna), Iwahig (Palawan), Mt Kitanglad (Bukidnon).

Paddyfield Pipit ■ *Anthus rufulus* 16.4cm

DESCRIPTION The default pipit across the Philippines. Upperparts light brown streaked with dark brown. Face has buff eye-stripe and thin dark brown whiskers. Wings dark brown with whitish edges that form two wing-bars. Breast buff coloured with fine streaking. Rest of underparts lighter buff colour, becoming darker at flanks. Bill black with lower bill yellowish. **DISTRIBUTION** Throughout the Philippines. **HABITS AND HABITATS** Usually seen foraging on the ground. Also observed standing erect in grass fields and rice fields, sprinting and pausing to stand alternately. Also seen in open lots and grassy fields. **SITES** Candaba Marsh (Pampanga), Mt Makiling (Laguna), Mt Kitanglad (Bukidnon).

Olive-backed Pipit ■ *Anthus hodgsoni* 14.4cm

DESCRIPTION Medium-sized pipit. Streaking can be seen on olive-coloured upperparts. Finer streaking on head. Breast buff coloured with heavy dark streaks. Flanks more faintly streaked and belly white. Buff eye-stripe can be seen, along with white spot behind eye in some birds. **DISTRIBUTION** Throughout the Philippines. **HABITS AND HABITATS** Seen in forests, usually with pine trees, in higher elevations. Walks continuously on the ground or on tree branches, foraging for worms. **SITES** Mt Polis (Mountain Province), Mt Kitanglad (Bukidnon).

White-cheeked Bullfinch ■ *Pyrrhula leucogenis* 16cm ⓔ

DESCRIPTION Medium-sized, chunky finch with white cheek and rump, and distinct silver bill. Top of head dark bluish-black; lores black, extending towards chin. Wings and uppertail-coverts dark bluish-black; upper back light brown. Greyish-brown wing-bar formed by greater wing-coverts. Belly greyish-white and undertail-coverts orange-buff. **DISTRIBUTION** Luzon, Panay, Mindanao. **HABITS AND HABITATS** Inhabits montane forests and forest edges but is more common in mossy forests, above 1,250m. Forages noisily, alone or joining mixed flocks. **SITES** Mt Polis (Mountain Province), Mt Kitanglad (Bukidnon).

Adult, Luzon

STATUS

A Accidental
NE Near Endemic
E Endemic
EX Extirpated
M Migrant
R Resident
SU Status Unknown

IUCN RED LIST STATUS
Conservation status is not given for species that have not yet been evaluated by the IUCN.

CR Critically Endangered
EN Endangered
VU Vulnerable
NT Near Threatened
DD Data Deficient

Common Name	Scientific Name	Status	IUCN
Megapodes	**Megapodiidae**		
Philippine Megapode	*Megapodius cumingii*	R	
Pheasants and allies	**Phasianidae**		
Chinese Francolin	*Francolinus pintadeanus*	I	
Daurian Partridge	*Perdix dauurica*	I	
Japanese Quail	*Coturnix japonica*	A	NT
King Quail	*Excalfactoria chinensis*	R	
Red Junglefowl	*Gallus gallus*	R	
Palawan Peacock-Pheasant	*Polyplectron napoleonis*	E	VU
Ducks, Geese and Swans	**Anatidae**		
Spotted Whistling Duck	*Dendrocygna guttata*	SU	
Wandering Whistling Duck	*Dendrocygna arcuata*	R	
Brant Goose	*Branta bernicla*	A	
Bar-headed Goose	*Anser indicus*	A	
Taiga Bean Goose	*Anser fabalis*	A	
Tundra Bean Goose	*Anser serrirostris*	A	
Greater White-fronted Goose	*Anser albifrons*	A	
Tundra Swan	*Cygnus columbianus*	A	
Common Shelduck	*Tadorna tadorna*	A	
Ruddy Shelduck	*Tadorna ferruginea*	A	
Mandarin Duck	*Aix galericulata*	A	
Cotton Pygmy Goose	*Nettapus coromandelianus*	A	
Baikal Teal	*Sibirionetta formosa*	A	
Garganey	*Spatula querquedula*	M	
Northern Shoveler	*Spatula clypeata*	M	
Gadwall	*Mareca strepera*	A	
Falcated Duck	*Mareca falcata*	A	NT
Eurasian Wigeon	*Mareca penelope*	M	
Philippine Duck	*Anas luzonica*	E	VU
Eastern Spot-billed Duck	*Anas zonorhyncha*	A	
Mallard	*Anas platyrhynchos*	A	

Northern Pintail	*Anas acuta*	M	
Eurasian Teal	*Anas crecca*	M	
Common Pochard	*Aythya ferina*	M	VU
Baer's Pochard	*Aythya baeri*	A	CR
Ferruginous Duck	*Aythya nyroca*	A	NT
Tufted Duck	*Aythya fuligula*	M	
Greater Scaup	*Aythya marila*	A	
Scaly-sided Merganser	*Mergus squamatus*	A	EN
Frogmouths	**Podargidae**		
Philippine Frogmouth	*Batrachostomus septimus*	E	
Palawan Frogmouth	*Batrachostomus chaseni*	E	
Nightjars	**Caprimulgidae**		
Great Eared Nightjar	*Lyncornis macrotis*	R	
Grey Nightjar	*Caprimulgus jotaka*	M	
Large-tailed Nightjar	*Caprimulgus macrurus*	R	
Philippine Nightjar	*Caprimulgus manillensis*	E	
Savanna Nightjar	*Caprimulgus affinis*	R	
Treeswifts	**Hemiprocnidae**		
Grey-rumped Treeswift	*Hemiprocne longipennis*	R	
Whiskered Treeswift	*Hemiprocne comata*	R	
Swifts	**Apodidae**		
Grey-rumped Swiftlet	*Collocalia marginata*	E	
Ridgetop Swiftlet	*Collocalia isonota*	E	
Pygmy Swiftlet	*Collocalia troglodytes*	E	
Philippine Swiftlet	*Aerodramus mearnsi*	E	
Whitehead's Swiftlet	*Aerodramus whiteheadi*	E	DD
Mossy-nest Swiftlet	*Aerodramus salangana*	SU	
Ameline Swiftlet	*Aerodramus amelis*	E	
Black-nest Swiftlet	*Aerodramus maximus*	SU	
Germain's Swiftlet	*Aerodramus germani*	R	
Philippine Spine-tailed Swift	*Mearnsia picina*	E	NT
White-throated Needletail	*Hirundapus caudacutus*	A	
Brown-backed Needletail	*Hirundapus giganteus*	R	
Purple Needletail	*Hirundapus celebensis*	R	
Asian Palm Swift	*Cypsiurus balasiensis*	R	
Pacific Swift	*Apus pacificus*	M (R?)	
House Swift	*Apus nipalensis*	R	
Cuckoos	**Cuculidae**		
Rufous Coucal	*Centropus unirufus*	E	NT
Black-faced Coucal	*Centropus melanops*	E	
Black-hooded Coucal	*Centropus steerii*	E	CR
Greater Coucal	*Centropus sinensis*	R	
Philippine Coucal	*Centropus viridis*	E	
Lesser Coucal	*Centropus bengalensis*	R	
Chestnut-breasted Malkoha	*Phaenicophaeus curvirostris*	R	
Rough-crested Malkoha	*Dasylophus superciliosus*	E	
Scale-feathered Malkoha	*Dasylophus cumingi*	E	
Chestnut-winged Cuckoo	*Clamator coromandus*	M	
Jacobin Cuckoo	*Clamator jacobinus*	A	
Asian Koel	*Eudynamys scolopaceus*	R	
Channel-billed Cuckoo	*Scythrops novaehollandiae*	A	
Violet Cuckoo	*Chrysococcyx xanthorhynchus*	R	
Little Bronze Cuckoo	*Chrysococcyx minutillus*	R	
Banded Bay Cuckoo	*Cacomantis sonneratii*	M (R?)	
Plaintive Cuckoo	*Cacomantis merulinus*	R	
Rusty-breasted Cuckoo	*Cacomantis sepulcralis*	R	
Philippine Drongo-Cuckoo	*Surniculus velutinus*	E	
Square-tailed Drongo-Cuckoo	*Surniculus lugubris*	R	
Large Hawk-Cuckoo	*Hierococcyx sparverioides*	M	
Philippine Hawk-Cuckoo	*Hierococcyx pectoralis*	E	

Indian Cuckoo	*Cuculus micropterus*	M	
Himalayan Cuckoo	*Cuculus saturatus*	M	
Oriental Cuckoo	*Cuculus optatus*	M	
Pigeons and Doves	**Columbidae**		
Rock Dove	*Columba livia*	I	
Metallic Pigeon	*Columba vitiensis*	R	
Oriental Turtle Dove	*Streptopelia orientalis*	A	
Island Collared Dove	*Streptopelia bitorquata*	R	VU
Red Turtle Dove	*Streptopelia tranquebarica*	R	
Spotted Dove	*Spilopelia chinensis*	R	
Philippine Cuckoo-Dove	*Macropygia tenuirostris*	NE	
Common Emerald Dove	*Chalcophaps indica*	R	
Zebra Dove	*Geopelia striata*	R	
Nicobar Pigeon	*Caloenas nicobarica*	R	NT
Luzon Bleeding-heart	*Gallicolumba luzonica*	E	NT
Mindanao Bleeding-heart	*Gallicolumba crinigera*	E	VU
Mindoro Bleeding-heart	*Gallicolumba platenae*	E	CR
Negros Bleeding-heart	*Gallicolumba keayi*	E	CR
Sulu Bleeding-heart	*Gallicolumba menagei*	E	CR
White-eared Brown Dove	*Phapitreron leucotis*	E	
Amethyst Brown Dove	*Phapitreron amethystinus*	E	
Tawitawi Brown Dove	*Phapitreron cinereiceps*	E	EN
Mindanao Brown Dove	*Phapitreron brunneiceps*	E	VU
Pink-necked Green Pigeon	*Treron vernans*	R	
Philippine Green Pigeon	*Treron axillaris*	E	
Thick-billed Green Pigeon	*Treron curvirostra*	R	
Whistling Green Pigeon	*Treron formosae*	R	NT
Flame-breasted Fruit Dove	*Ptilinopus marchei*	E	VU
Cream-breasted Fruit Dove	*Ptilinopus merrilli*	E	NT
Yellow-breasted Fruit Dove	*Ptilinopus occipitalis*	E	
Black-chinned Fruit Dove	*Ptilinopus leclancheri*	NE	
Superb Fruit Dove	*Ptilinopus superbus*	SU	
Black-naped Fruit Dove	*Ptilinopus melanospilus*	R	
Negros Fruit Dove	*Ptilinopus arcanus*	E	CR
Pink-bellied Imperial Pigeon	*Ducula poliocephala*	E	NT
Mindoro Imperial Pigeon	*Ducula mindorensis*	E	EN
Spotted Imperial Pigeon	*Ducula carola*	E	VU
Green Imperial Pigeon	*Ducula aenea*	R	
Grey Imperial Pigeon	*Ducula pickeringii*	NE	VU
Pied Imperial Pigeon	*Ducula bicolor*	R	
Crakes, Rails and Coots	**Rallidae**		
Red-legged Crake	*Rallina fasciata*	SU	
Slaty-legged Crake	*Rallina eurizonoides*	R	
Calayan Rail	*Gallirallus calayanensis*	E	VU
Barred Rail	*Gallirallus torquatus*	R	
Buff-banded Rail	*Gallirallus philippensis*	R	
Slaty-breasted Rail	*Gallirallus striatus*	R	
Brown-banded Rail	*Lewinia mirifica*	E	DD
Plain Bush-hen	*Amaurornis olivacea*	E	
White-breasted Waterhen	*Amaurornis phoenicurus*	R	
Baillon's Crake	*Porzana pusilla*	M	
Ruddy-breasted Crake	*Porzana fusca*	R	
Spotless Crake	*Porzana tabuensis*	R	
White-browed Crake	*Porzana cinerea*	R	
Watercock	*Gallicrex cinerea*	R	
Philippine Swamphen	*Porphyrio pulverulentus*	E	
Common Moorhen	*Gallinula chloropus*	R,M	
Eurasian Coot	*Fulica atra*	M	
Cranes	**Gruidae**		
Sarus Crane	*Antigone antigone*	EX	VU

Demoiselle Crane	*Grus virgo*	A	
Hooded Crane	*Grus monacha*	A	VU
Grebes	**Podicipedidae**		
Little Grebe	*Tachybaptus ruficollis*	R	
Black-necked Grebe	*Podiceps nigricollis*	A	
Buttonquails	**Turnicidae**		
Common Buttonquail	*Turnix sylvaticus*	R	
Spotted Buttonquail	*Turnix ocellatus*	E	
Barred Buttonquail	*Turnix suscitator*	R	
Worcester's Buttonquail	*Turnix worcesteri*	E	DD
Stone-curlews, Thick-knees	**Burhinidae**		
Beach Stone-curlew	*Esacus magnirostris*	R	NT
Oystercatchers	**Haematopodidae**		
Eurasian Oystercatcher	*Haematopus ostralegus*	A	NT
Stilts and Avocets	**Recurvirostridae**		
Black-winged Stilt	*Himantopus himantopus*	M (R?)	
Pied Stilt	*Himantopus leucocephalus*	R,M	
Pied Avocet	*Recurvirostra avosetta*	A	
Plovers	**Charadriidae**		
Northern Lapwing	*Vanellus vanellus*	A	NT
Grey-headed Lapwing	*Vanellus cinereus*	A	
Pacific Golden Plover	*Pluvialis fulva*	M	
Grey Plover	*Pluvialis squatarola*	M	
Common Ringed Plover	*Charadrius hiaticula*	A	
Little Ringed Plover	*Charadrius dubius*	R,M	
Kentish Plover	*Charadrius alexandrinus*	M	
Malaysian Plover	*Charadrius peronii*	R	NT
Lesser Sand Plover	*Charadrius mongolus*	M	
Greater Sand Plover	*Charadrius leschenaultii*	M	
Oriental Plover	*Charadrius veredus*	A	
Painted-snipes	**Rostratulidae**		
Greater Painted-snipe	*Rostratula benghalensis*	R	
Jacanas	**Jacanidae**		
Comb-crested Jacana	*Irediparra gallinacea*	R	
Pheasant-tailed Jacana	*Hydrophasianus chirurgus*	R	
Sandpipers and Snipes	**Scolopacidae**		
Bristle-thighed Curlew	*Numenius tahitiensis*	A	VU
Eurasian Whimbrel	*Numenius phaeopus*	M	
Little Curlew	*Numenius minutus*	M	
Far Eastern Curlew	*Numenius madagascariensis*	M	EN
Eurasian Curlew	*Numenius arquata*	M	NT
Bar-tailed Godwit	*Limosa lapponica*	M	NT
Black-tailed Godwit	*Limosa limosa*	M	NT
Ruddy Turnstone	*Arenaria interpres*	M	
Great Knot	*Calidris tenuirostris*	M	EN
Red Knot	*Calidris canutus*	M	NT
Ruff	*Calidris pugnax*	M	
Broad-billed Sandpiper	*Calidris falcinellus*	M	
Sharp-tailed Sandpiper	*Calidris acuminata*	M	
Curlew Sandpiper	*Calidris ferruginea*	M	NT
Temminck's Stint	*Calidris temminckii*	A	
Long-toed Stint	*Calidris subminuta*	M	
Red-necked Stint	*Calidris ruficollis*	M	NT
Sanderling	*Calidris alba*	M	
Dunlin	*Calidris alpina*	A	
Little Stint	*Calidris minuta*	A	
Pectoral Sandpiper	*Calidris melanotos*	A	
Asian Dowitcher	*Limnodromus semipalmatus*	M	NT
Long-billed Dowitcher	*Limnodromus scolopaceus*	A	
Eurasian Woodcock	*Scolopax rusticola*	A	

Bukidnon Woodcock	*Scolopax bukidnonensis*	E	
Jack Snipe	*Lymnocryptes minimus*	A	
Latham's Snipe	*Gallinago hardwickii*	A	
Pin-tailed Snipe	*Gallinago stenura*	M	
Swinhoe's Snipe	*Gallinago megala*	M	
Common Snipe	*Gallinago gallinago*	M	
Terek Sandpiper	*Xenus cinereus*	M	
Red-necked Phalarope	*Phalaropus lobatus*	M	
Red Phalarope	*Phalaropus fulicarius*	A	
Common Sandpiper	*Actitis hypoleucos*	M	
Green Sandpiper	*Tringa ochropus*	M	
Grey-tailed Tattler	*Tringa brevipes*	M	NT
Common Redshank	*Tringa totanus*	M	
Marsh Sandpiper	*Tringa stagnatilis*	M	
Wood Sandpiper	*Tringa glareola*	M	
Spotted Redshank	*Tringa erythropus*	A	
Common Greenshank	*Tringa nebularia*	M	
Nordmann's Greenshank	*Tringa guttifer*	A	EN
Pratincoles	**Glareolidae**		
Oriental Pratincole	*Glareola maldivarum*	R,M	
Gulls and Terns	**Laridae**		
Brown Noddy	*Anous stolidus*	R	
Black Noddy	*Anous minutus*	R	
White Tern	*Gygis alba*	A	
Black-headed Gull	*Chroicocephalus ridibundus*	M	
Saunders's Gull	*Chroicocephalus saundersi*	A	VU
Laughing Gull	*Leucophaeus atricilla*	A	
Franklin's Gull	*Leucophaeus pipixcan*	A	
Black-tailed Gull	*Larus crassirostris*	A	
Mew Gull	*Larus canus*	A	
Vega Gull	*Larus vegae*	A	
Slaty-backed Gull	*Larus schistisagus*	A	
Lesser Black-backed Gull	*Larus fuscus*	A	
Gull-billed Tern	*Gelochelidon nilotica*	M	
Australian Tern	*Gelochelidon macrotarsa*	A	
Caspian Tern	*Hydroprogne caspia*	M	
Greater Crested Tern	*Thalasseus bergii*	R	
Chinese Crested Tern	*Thalasseus bernsteini*	A	CR
Little Tern	*Sternula albifrons*	R,M	
Aleutian Tern	*Onychoprion aleuticus*	A	VU
Bridled Tern	*Onychoprion anaethetus*	R	
Sooty Tern	*Onychoprion fuscatus*	R,M	
Roseate Tern	*Sterna dougallii*	M (R?)	
Black-naped Tern	*Sterna sumatrana*	R	
Common Tern	*Sterna hirundo*	R	
Whiskered Tern	*Chlidonias hybrida*	M	
White-winged Tern	*Chlidonias leucopterus*	M	
Skuas	**Stercorariidae**		
Pomarine Jaeger	*Stercorarius pomarinus*	M	
Parasitic Jaeger	*Stercorarius parasiticus*	A	
Long-tailed Jaeger	*Stercorarius longicaudus*	A	
Tropicbirds	**Phaethontidae**		
Red-tailed Tropicbird	*Phaethon rubricauda*	A	
White-tailed Tropicbird	*Phaethon lepturus*	A	
Albatrosses	**Diomedeidae**		
Laysan Albatross	*Phoebastria immutabilis*	A	NT
Northern Storm Petrels	**Hydrobatidae**		
Swinhoe's Storm Petrel	*Oceanodroma monorhis*	A	NT
Leach's Storm Petrel	*Oceanodroma leucorhoa*	A	VU

Petrels and Shearwaters	Procellariidae		
Kermadec Petrel	*Pterodroma neglecta*	A	
Hawaiian Petrel	*Pterodroma sandwichensis*	A	EN
Bonin Petrel	*Pterodroma hypoleuca*	A	
Black-winged Petrel	*Pterodroma nigripennis*	A	
Tahiti Petrel	*Pseudobulweria rostrata*	A	NT
Streaked Shearwater	*Calonectris leucomelas*	M	NT
Wedge-tailed Shearwater	*Ardenna pacifica*	M	
Short-tailed Shearwater	*Ardenna tenuirostris*	A	
Bulwer's Petrel	*Bulweria bulwerii*	A	
Storks	**Ciconiidae**		
Black Stork	*Ciconia nigra*	A	
Woolly-necked Stork	*Ciconia episcopus*	R	VU
Oriental Stork	*Ciconia boyciana*	A	EN
Frigatebirds	**Fregatidae**		
Christmas Frigatebird	*Fregata andrewsi*	M	CR
Great Frigatebird	*Fregata minor*	M	
Lesser Frigatebird	*Fregata ariel*	M	
Boobies	**Sulidae**		
Masked Booby	*Sula dactylatra*	R	
Red-footed Booby	*Sula sula*	R	
Brown Booby	*Sula leucogaster*	R,M	
Cormorants	**Phalacrocoracidae**		
Great Cormorant	*Phalacrocorax carbo*	M	
Darters	**Anhingidae**		
Oriental Darter	*Anhinga melanogaster*	R	NT
Ibises and Spoonbills	**Threskiornithidae**		
Black-headed Ibis	*Threskiornis melanocephalus*	A	NT
Glossy Ibis	*Plegadis falcinellus*	R	
Eurasian Spoonbill	*Platalea leucorodia*	A	
Black-faced Spoonbill	*Platalea minor*	A	EN
Bitterns, Egrets and Herons	**Ardeidae**		
Eurasian Bittern	*Botaurus stellaris*	A	
Yellow Bittern	*Ixobrychus sinensis*	R	
Von Schrenck's Bittern	*Ixobrychus eurhythmus*	M	
Cinnamon Bittern	*Ixobrychus cinnamomeus*	R	
Black Bittern	*Ixobrychus flavicollis*	R	
Japanese Night Heron	*Gorsachius goisagi*	M	EN
Malayan Night Heron	*Gorsachius melanolophus*	R	
Black-crowned Night Heron	*Nycticorax nycticorax*	R	
Nankeen Night Heron	*Nycticorax caledonicus*	R	
Striated Heron	*Butorides striata*	R,M	
Chinese Pond Heron	*Ardeola bacchus*	M	
Javan Pond Heron	*Ardeola speciosa*	R	
Eastern Cattle Egret	*Bubulcus coromandus*	R,M	
Grey Heron	*Ardea cinerea*	M	
Great-billed Heron	*Ardea sumatrana*	R	
Purple Heron	*Ardea purpurea*	R	
Great Egret	*Ardea alba*	R,M	
Intermediate Egret	*Ardea intermedia*	R,M	
Little Egret	*Egretta garzetta*	R,M	
Pacific Reef Heron	*Egretta sacra*	R	
Chinese Egret	*Egretta eulophotes*	M	VU
Pelicans	**Pelecanidae**		
Spot-billed Pelican	*Pelecanus philippensis*	EX	NT
Dalmatian Pelican	*Pelecanus crispus*	A	NT
Australian Pelican	*Pelecanus conspicillatus*	A	
Ospreys	**Pandionidae**		
Western Osprey	*Pandion haliaetus*	M (R?)	

Kites, Hawks and Eagles	**Accipitridae**		
Black-winged Kite	*Elanus caeruleus*	R	
Crested Honey Buzzard	*Pernis ptilorhynchus*	R,M	
Philippine Honey Buzzard	*Pernis steerei*	E	
Jerdon's Baza	*Aviceda jerdoni*	R (M?)	
Cinereous Vulture	*Aegypius monachus*	A	NT
Crested Serpent Eagle	*Spilornis cheela*	R	
Philippine Serpent Eagle	*Spilornis holospilus*	E	
Philippine Eagle	*Pithecophaga jefferyi*	E	CR
Changeable Hawk-Eagle	*Nisaetus cirrhatus*	R	
Philippine Hawk-Eagle	*Nisaetus philippensis*	E	EN
Pinsker's Hawk-Eagle	*Nisaetus pinskeri*	E	EN
Rufous-bellied Eagle	*Lophotriorchis kienerii*	R	NT
Crested Goshawk	*Accipiter trivirgatus*	R	
Shikra	*Accipiter badius*	A	
Chinese Sparrowhawk	*Accipiter soloensis*	M	
Japanese Sparrowhawk	*Accipiter gularis*	M	
Besra	*Accipiter virgatus*	R	
Eurasian Sparrowhawk	*Accipiter nisus*	A	
Eastern Marsh Harrier	*Circus spilonotus*	M	
Pied Harrier	*Circus melanoleucos*	R,M	
Black Kite	*Milvus migrans*	A	
Brahminy Kite	*Haliastur indus*	R	
White-bellied Sea Eagle	*Haliaeetus leucogaster*	R	
Grey-headed Fish Eagle	*Haliaeetus ichthyaetus*	R	NT
Grey-faced Buzzard	*Butastur indicus*	M	
Eastern Buzzard	*Buteo japonicus*	R (M?)	
Barn Owls	**Tytonidae**		
Eastern Grass Owl	*Tyto longimembris*	R	
Owls	**Strigidae**		
Giant Scops Owl	*Otus gurneyi*	E	VU
Palawan Scops Owl	*Otus fuliginosus*	E	NT
Philippine Scops Owl	*Otus megalotis*	E	
Everett's Scops Owl	*Otus everetti*	E	
Negros Scops Owl	*Otus nigrorum*	E	VU
Mindanao Scops Owl	*Otus mirus*	E	NT
Luzon Scops Owl	*Otus longicornis*	E	NT
Mindoro Scops Owl	*Otus mindorensis*	E	NT
Oriental Scops Owl	*Otus sunia*	A	
Mantanani Scops Owl	*Otus mantananensis*	NE	NT
Ryukyu Scops Owl	*Otus elegans*	R	NT
Philippine Eagle-Owl	*Bubo philippensis*	E	VU
Spotted Wood Owl	*Strix seloputo*	R	
Brown Hawk-Owl	*Ninox scutulata*	R	
Northern Boobook	*Ninox japonica*	R,M	
Chocolate Boobook	*Ninox randi*	NE	NT
Luzon Hawk-Owl	*Ninox philippensis*	E	
Mindanao Hawk-Owl	*Ninox spilocephala*	E	NT
Mindoro Hawk-Owl	*Ninox mindorensis*	E	VU
Romblon Hawk-Owl	*Ninox spilonotus*	E	EN
Cebu Hawk-Owl	*Ninox rumseyi*	E	EN
Camiguin Hawk-Owl	*Ninox leventisi*	E	EN
Sulu Hawk-Owl	*Ninox reyi*	E	VU
Short-eared Owl	*Asio flammeus*	A	
Trogons	**Trogonidae**		
Philippine Trogon	*Harpactes ardens*	E	
Hoopoes	**Upupidae**		
Eurasian Hoopoe	*Upupa epops*	A	
Hornbills	**Bucerotidae**		
Rufous Hornbill	*Buceros hydrocorax*	E	VU

Palawan Hornbill	*Anthracoceros marchei*	E	VU
Sulu Hornbill	*Anthracoceros montani*	E	CR
Walden's Hornbill	*Rhabdotorrhinus waldeni*	E	CR
Writhed Hornbill	*Rhabdotorrhinus leucocephalus*	E	NT
Luzon Hornbill	*Penelopides manillae*	E	
Mindoro Hornbill	*Penelopides mindorensis*	E	EN
Mindanao Hornbill	*Penelopides affinis*	E	
Samar Hornbill	*Penelopides samarensis*	E	
Visayan Hornbill	*Penelopides panini*	E	EN
Rollers	**Coraciidae**		
Oriental Dollarbird	*Eurystomus orientalis*	R	
Kingfishers	**Alcedinidae**		
Spotted Wood Kingfisher	*Actenoides lindsayi*	E	
Hombron's Kingfisher	*Actenoides hombroni*	E	VU
Stork-billed Kingfisher	*Pelargopsis capensis*	R	
Ruddy Kingfisher	*Halcyon coromanda*	R,M	
White-throated Kingfisher	*Halcyon smyrnensis*	R	
Black-capped Kingfisher	*Halcyon pileata*	R	
Winchell's Kingfisher	*Todiramphus winchelli*	E	VU
Collared Kingfisher	*Todiramphus chloris*	R	
Sacred Kingfisher	*Todiramphus sanctus*	A	
Blue-eared Kingfisher	*Alcedo meninting*	R	
Common Kingfisher	*Alcedo atthis*	M	
Oriental Dwarf Kingfisher	*Ceyx erithaca*	R	
Philippine Dwarf Kingfisher	*Ceyx melanurus*	E	VU
Dimorphic Dwarf Kingfisher	*Ceyx margarethae*	E	
Indigo-banded Kingfisher	*Ceyx cyanopectus*	E	
Southern Silvery Kingfisher	*Ceyx argentatus*	E	NT
Northern Silvery Kingfisher	*Ceyx flumenicola*	E	NT
Bee-eaters	**Meropidae**		
Blue-tailed Bee-eater	*Merops philippinus*	R	
Blue-throated Bee-eater	*Merops viridis*	R	
Asian Barbets	**Megalaimidae**		
Coppersmith Barbet	*Psilopogon haemacephalus*	R	
Woodpeckers	**Picidae**		
Philippine Pygmy Woodpecker	*Yungipicus maculatus*	E	
Sulu Pygmy Woodpecker	*Yungipicus ramsayi*	E	VU
White-bellied Woodpecker	*Dryocopus javensis*	R	
Spot-throated Flameback	*Dinopium everetti*	E	NT
Buff-spotted Flameback	*Chrysocolaptes lucidus*	E	
Luzon Flameback	*Chrysocolaptes haematribon*	E	
Yellow-faced Flameback	*Chrysocolaptes xanthocephalus*	E	EN
Red-headed Flameback	*Chrysocolaptes erythrocephalus*	E	EN
Sooty Woodpecker	*Mulleripicus funebris*	E	NT
Great Slaty Woodpecker	*Mulleripicus pulverulentus*	R	VU
Falconets and Falcons	**Falconidae**		
Philippine Falconet	*Microhierax erythrogenys*	E	
Common Kestrel	*Falco tinnunculus*	M	
Spotted Kestrel	*Falco moluccensis*	R	
Amur Falcon	*Falco amurensis*	A	
Merlin	*Falco columbarius*	A	
Eurasian Hobby	*Falco subbuteo*	A	
Oriental Hobby	*Falco severus*	R	
Peregrine Falcon	*Falco peregrinus*	R,M	
Cockatoos	**Cacatuidae**		
Red-vented Cockatoo	*Cacatua haematuropygia*	E	CR
Old World Parrots	**Psittaculidae**		
Mindanao Racket-tail	*Prioniturus waterstradti*	E	NT
Montane Racket-tail	*Prioniturus montanus*	E	NT
Blue-headed Racket-tail	*Prioniturus platenae*	E	VU

Mindoro Racket-tail	*Prioniturus mindorensis*	E	VU
Blue-winged Racket-tail	*Prioniturus verticalis*	E	CR
Green Racket-tail	*Prioniturus luconensis*	E	EN
Blue-crowned Racket-tail	*Prioniturus discurus*	E	
Great-billed Parrot	*Tanygnathus megalorynchos*	R	
Blue-naped Parrot	*Tanygnathus lucionensis*	NE	NT
Blue-backed Parrot	*Tanygnathus sumatranus*	R	
Rose-ringed Parakeet	*Psittacula krameri*	I	
Mindanao Lorikeet	*Trichoglossus johnstoniae*	E	NT
Guaiabero	*Bolbopsittacus lunulatus*	E	
Philippine Hanging Parrot	*Loriculus philippensis*	E	
Camiguin Hanging Parrot	*Loriculus camiguinensis*	E	NE
Broadbills	**Eurylaimidae**		
Wattled Broadbill	*Sarcophanops steerii*	E	VU
Visayan Broadbill	*Sarcophanops samarensis*	E	VU
Pittas	**Pittidae**		
Whiskered Pitta	*Erythropitta kochi*	E	NT
Philippine Pitta	*Erythropitta erythrogaster*	NE	
Hooded Pitta	*Pitta sordida*	R	
Azure-breasted Pitta	*Pitta steerii*	E	VU
Fairy Pitta	*Pitta nympha*	A	VU
Blue-winged Pitta	*Pitta moluccensis*	A	
Australasian Warblers	**Acanthizidae**		
Golden-bellied Gerygone	*Gerygone sulphurea*	R	
Woodswallows	**Artamidae**		
White-breasted Woodswallow	*Artamus leucorynchus*	R	
Ioras	**Aegithinidae**		
Common Iora	*Aegithina tiphia*	R	
Cuckooshrikes	**Campephagidae**		
Fiery Minivet	*Pericrocotus igneus*	R	NT
Scarlet Minivet	*Pericrocotus speciosus*	R	
Ashy Minivet	*Pericrocotus divaricatus*	M	
Bar-bellied Cuckooshrike	*Coracina striata*	R	
McGregor's Cuckooshrike	*Malindangia mcgregori*	E	
White-winged Cuckooshrike	*Edolisoma ostentum*	E	VU
Blackish Cuckooshrike	*Edolisoma coerulescens*	E	
Black-bibbed Cicadabird	*Edolisoma mindanense*	E	VU
Black-and-white Triller	*Lalage melanoleuca*	E	
Pied Triller	*Lalage nigra*	R	
Black-winged Cuckooshrike	*Lalage melaschistos*	A	
Whistlers	**Pachycephalidae**		
Mangrove Whistler	*Pachycephala cinerea*	R	
Green-backed Whistler	*Pachycephala albiventris*	E	
White-vented Whistler	*Pachycephala homeyeri*	NE	
Yellow-bellied Whistler	*Pachycephala philippinensis*	E	
Shrikes	**Laniidae**		
Tiger Shrike	*Lanius tigrinus*	A	
Brown Shrike	*Lanius cristatus*	M	
Long-tailed Shrike	*Lanius schach*	R	
Mountain Shrike	*Lanius validirostris*	E	NT
Orioles	**Oriolidae**		
Isabela Oriole	*Oriolus isabellae*	E	CR
Dark-throated Oriole	*Oriolus xanthonotus*	R	NT
Philippine Oriole	*Oriolus steerii*	E	
White-lored Oriole	*Oriolus albiloris*	E	
Black-naped Oriole	*Oriolus chinensis*	R	
Drongos	**Dicruridae**		
Black Drongo	*Dicrurus macrocercus*	A	
Ashy Drongo	*Dicrurus leucophaeus*	R,M	
Crow-billed Drongo	*Dicrurus annectens*	A	

Balicassiao	*Dicrurus balicassius*	E	
Hair-crested Drongo	*Dicrurus hottentottus*	R	
Tablas Drongo	*Dicrurus menagei*	E	EN
Fantails	**Rhipiduridae**		
Mindanao Blue Fantail	*Rhipidura superciliaris*	E	
Visayan Blue Fantail	*Rhipidura samarensis*	E	
Blue-headed Fantail	*Rhipidura cyaniceps*	E	
Tablas Fantail	*Rhipidura sauli*	E	VU
Visayan Fantail	*Rhipidura albiventris*	E	
Philippine Pied Fantail	*Rhipidura nigritorquis*	E	
Black-and-cinnamon Fantail	*Rhipidura nigrocinnamomea*	E	
Monarchs	**Monarchidae**		
Black-naped Monarch	*Hypothymis azurea*	R	
Short-crested Monarch	*Hypothymis helenae*	E	NT
Celestial Monarch	*Hypothymis coelestis*	E	VU
Amur Paradise Flycatcher	*Terpsiphone incei*	A	
Japanese Paradise Flycatcher	*Terpsiphone atrocaudata*	R,M	NT
Blue Paradise Flycatcher	*Terpsiphone cyanescens*	E	
Rufous Paradise Flycatcher	*Terpsiphone cinnamomea*	NE	
Crows	**Corvidae**		
Slender-billed Crow	*Corvus enca*	R	
Large-billed Crow	*Corvus macrorhynchos*	R	
Waxwings	**Bombycillidae**		
Japanese Waxwing	*Bombycilla japonica*	A	NT
Fairy Flycatchers	**Stenostiridae**		
Citrine Canary-flycatcher	*Culicicapa helianthea*	R	
Tits	**Paridae**		
Elegant Tit	*Pardaliparus elegans*	E	
Palawan Tit	*Pardaliparus amabilis*	E	NT
White-fronted Tit	*Sittiparus semilarvatus*	E	NT
Larks	**Alaudidae**		
Horsfield's Bush Lark	*Mirafra javanica*	R	
Oriental Skylark	*Alauda gulgula*	R	
Bulbuls	**Pycnonotidae**		
Black-headed Bulbul	*Pycnonotus atriceps*	R	
Light-vented Bulbul	*Pycnonotus sinensis*	A	
Yellow-wattled Bulbul	*Pycnonotus urostictus*	E	
Yellow-vented Bulbul	*Pycnonotus goiavier*	R	
Olive-winged Bulbul	*Pycnonotus plumosus*	R	
Ashy-fronted Bulbul	*Pycnonotus cinereifrons*	E	
Palawan Bulbul	*Alophoixus frater*	E	
Sulphur-bellied Bulbul	*Iole palawanensis*	E	
Black Bulbul	*Hypsipetes leucocephalus*	A	
Philippine Bulbul	*Hypsipetes philippinus*	E	
Mindoro Bulbul	*Hypsipetes mindorensis*	E	
Visayan Bulbul	*Hypsipetes guimarasensis*	E	
Zamboanga Bulbul	*Hypsipetes rufigularis*	E	NT
Streak-breasted Bulbul	*Hypsipetes siquijorensis*	E	EN
Yellowish Bulbul	*Hypsipetes everetti*	E	
Brown-eared Bulbul	*Hypsipetes amaurotis*	R	
Swallows and Martins	**Hirundinidae**		
Grey-throated Martin	*Riparia chinensis*	R	
Sand Martin	*Riparia riparia*	M	
Barn Swallow	*Hirundo rustica*	M	
Pacific Swallow	*Hirundo tahitica*	R	
Asian House Martin	*Delichon dasypus*	A	
Striated Swallow	*Cecropis striolata*	R	
Cettia Bush Warblers and allies	**Cettiidae**		
Mountain Tailorbird	*Phyllergates cucullatus*	R	
Rufous-headed Tailorbird	*Phyllergates heterolaemus*	E	

Philippine Bush Warbler	*Horornis seebohmi*	E	
Manchurian Bush Warbler	*Horornis canturians*	M	
Sunda Bush Warbler	*Horornis vulcanius*	R	
Asian Stubtail	*Urosphena squameiceps*	A	
Leaf Warblers and allies	**Phylloscopidae**		
Yellow-browed Warbler	*Phylloscopus inornatus*	A	
Dusky Warbler	*Phylloscopus fuscatus*	A	
Willow Warbler	*Phylloscopus trochilus*	A	
Ijima's Leaf Warbler	*Phylloscopus ijimae*	A	VU
Philippine Leaf Warbler	*Phylloscopus olivaceus*	E	
Lemon-throated Leaf Warbler	*Phylloscopus cebuensis*	E	
Japanese Leaf Warbler	*Phylloscopus xanthodryas*	M	
Kamchatka Leaf Warbler	*Phylloscopus examinandus*	M	
Arctic Warbler	*Phylloscopus borealis*	M	
Yellow-breasted Warbler	*Phylloscopus montis*	R	
Negros Leaf Warbler	*Phylloscopus nigrorum*	E	
Reed Warblers and allies	**Acrocephalidae**		
Oriental Reed Warbler	*Acrocephalus orientalis*	M	
Clamorous Reed Warbler	*Acrocephalus stentoreus*	R	
Black-browed Reed Warbler	*Acrocephalus bistrigiceps*	A	
Speckled Reed Warbler	*Acrocephalus sorghophilus*	M	EN
Grassbirds and allies	**Locustellidae**		
Cordillera Ground Warbler	*Robsonius rabori*	E	VU
Sierra Madre Ground Warbler	*Robsonius thompsoni*	E	
Bicol Ground Warbler	*Robsonius sorsogonensis*	E	NT
Gray's Grasshopper Warbler	*Helopsaltes fasciolatus*	M	
Pallas's Grasshopper Warbler	*Helopsaltes certhiola*	A	
Middendorff's Grasshopper Warbler	*Helopsaltes ochotensis*	M	
Lanceolated Warbler	*Locustella lanceolata*	M	
Long-tailed Bush Warbler	*Locustella caudata*	E	
Benguet Bush Warbler	*Locustella seebohmi*	E	
Tawny Grassbird	*Cincloramphus timoriensis*	R	
Striated Grassbird	*Megalurus palustris*	R	
Cisticolas and allies	**Cisticolidae**		
Zitting Cisticola	*Cisticola juncidis*	R	
Golden-headed Cisticola	*Cisticola exilis*	R	
Cisticolas and allies	**Timaliidae**		
Visayan Miniature Babbler	*Micromacronus leytensis*	E	DD
Mindanao Miniature Babbler	*Micromacronus sordidus*	E	NT
Cisticolas and allies	**Cisticolidae**		
Philippine Tailorbird	*Orthotomus castaneiceps*	E	
Trilling Tailorbird	*Orthotomus chloronotus*	E	
Rufous-fronted Tailorbird	*Orthotomus frontalis*	E	
Grey-backed Tailorbird	*Orthotomus derbianus*	E	
Rufous-tailed Tailorbird	*Orthotomus sericeus*	R	
Ashy Tailorbird	*Orthotomus ruficeps*	R	
White-eared Tailorbird	*Orthotomus cinereiceps*	E	
Black-headed Tailorbird	*Orthotomus nigriceps*	E	
Yellow-breasted Tailorbird	*Orthotomus samarensis*	E	NT
Babblers	**Timaliidae**		
Pin-striped Tit-Babbler	*Macronus gularis*	R	
Bold-striped Tit-Babbler	*Macronus bornensis*	R	
Brown Tit-Babbler	*Macronus striaticeps*	E	
Ground Babblers	**Pellorneidae**		
Striated Wren-Babbler	*Ptilocichla mindanensis*	E	
Falcated Wren-Babbler	*Ptilocichla falcata*	E	VU
Ashy-headed Babbler	*Malacocincla cinereiceps*	E	
Melodious Babbler	*Malacopteron palawanense*	E	NT
White-eyes	**Zosteropidae**		
Chestnut-faced Babbler	*Zosterornis whiteheadi*	E	

Luzon Striped Babbler	*Zosterornis striatus*	E	NT
Panay Striped Babbler	*Zosterornis latistriatus*	E	NT
Negros Striped Babbler	*Zosterornis nigrorum*	E	EN
Palawan Striped Babbler	*Zosterornis hypogrammicus*	E	NT
Flame-templed Babbler	*Dasycrotapha speciosa*	E	EN
Mindanao Pygmy Babbler	*Dasycrotapha plateni*	E	NT
Visayan Pygmy Babbler	*Dasycrotapha pygmaea*	E	NT
Golden-crowned Babbler	*Sterrhoptilus dennistouni*	E	NT
Black-crowned Babbler	*Sterrhoptilus nigrocapitatus*	E	
Rusty-crowned Babbler	*Sterrhoptilus capitalis*	E	
Mindanao White-eye	*Lophozosterops goodfellowi*	E	
Warbling White-eye	*Zosterops japonicus*	R	
Lowland White-eye	*Zosterops meyeni*	NE	
Everett's White-eye	*Zosterops everetti*	NE	
Yellowish White-eye	*Zosterops nigrorum*	E	
Fairy-bluebirds	**Irenidae**		
Asian Fairy-bluebird	*Irena puella*	R	NT
Philippine Fairy-bluebird	*Irena cyanogastra*	E	NT
Nuthatches	**Sittidae**		
Velvet-fronted Nuthatch	*Sitta frontalis*	R	
Sulphur-billed Nuthatch	*Sitta oenochlamys*	E	
Starlings and Rhabdornis	**Sturnidae**		
Asian Glossy Starling	*Aplonis panayensis*	R	
Short-tailed Starling	*Aplonis minor*	R	
Apo Myna	*Basilornis mirandus*	E	NT
Coleto	*Sarcops calvus*	NE	
Common Hill Myna	*Gracula religiosa*	R	
Crested Myna	*Acridotheres cristatellus*	I	
Common Myna	*Acridotheres tristis*	A	
Red-billed Starling	*Spodiopsar sericeus*	A	
White-cheeked Starling	*Spodiopsar cineraceus*	A	
Daurian Starling	*Agropsar sturninus*	A	
Chestnut-cheeked Starling	*Agropsar philippensis*	M	
White-shouldered Starling	*Sturnia sinensis*	M	
Rosy Starling	*Pastor roseus*	A	
Common Starling	*Sturnus vulgaris*	A	
Stripe-headed Rhabdornis	*Rhabdornis mystacalis*	E	
Stripe-breasted Rhabdornis	*Rhabdornis inornatus*	E	
Grand Rhabdornis	*Rhabdornis grandis*	E	
Thrushes	**Turdidae**		
Chestnut-capped Thrush	*Geokichla interpres*	R	NT
Ashy Thrush	*Geokichla cinerea*	E	VU
Siberian Thrush	*Geokichla sibirica*	A	
Sunda Thrush	*Zoothera andromedae*	R	
White's Thrush	*Zoothera aurea*	M	
Grey-backed Thrush	*Turdus hortulorum*	A	
Chinese Blackbird	*Turdus mandarinus*	A	
Island Thrush	*Turdus poliocephalus*	R	
Eyebrowed Thrush	*Turdus obscurus*	M	
Pale Thrush	*Turdus pallidus*	A	
Brown-headed Thrush	*Turdus chrysolaus*	M	
Naumann's Thrush	*Turdus naumanni*	A	
Dusky Thrush	*Turdus eunomus*	A	
Chats and Old World Flycatchers	**Muscicapidae**		
Philippine Magpie-Robin	*Copsychus mindanensis*	E	
White-browed Shama	*Copsychus luzoniensis*	E	
White-vented Shama	*Copsychus niger*	E	
Black Shama	*Copsychus cebuensis*	E	EN
Grey-streaked Flycatcher	*Muscicapa griseisticta*	M	
Dark-sided Flycatcher	*Muscicapa sibirica*	M	

Asian Brown Flycatcher	*Muscicapa dauurica*	M	
Ashy-breasted Flycatcher	*Muscicapa randi*	E	VU
Ferruginous Flycatcher	*Muscicapa ferruginea*	M	
Blue-breasted Blue Flycatcher	*Cyornis herioti*	E	NT
Palawan Blue Flycatcher	*Cyornis lemprieri*	E	NT
Mangrove Blue Flycatcher	*Cyornis rufigastra*	R	
Rufous-tailed Jungle Flycatcher	*Cyornis ruficauda*	R	
Blue-and-white Flycatcher	*Cyanoptila cyanomelana*	M	
Verditer Flycatcher	*Eumyias thalassinus*	A	
Turquoise Flycatcher	*Eumyias panayensis*	R	
Bagobo Babbler	*Leonardina woodi*	E	
White-browed Shortwing	*Brachypteryx montana*	R	
White-throated Jungle Flycatcher	*Vauriella albigularis*	E	EN
White-browed Jungle Flycatcher	*Vauriella insignis*	E	VU
Slaty-backed Jungle Flycatcher	*Vauriella goodfellowi*	E	NT
Siberian Blue Robin	*Larvivora cyane*	A	
Bluethroat	*Luscinia svecica*	A	
Siberian Rubythroat	*Calliope calliope*	M	
Red-flanked Bluetail	*Tarsiger cyanurus*	A	
Yellow-rumped Flycatcher	*Ficedula zanthopygia*	A	
Narcissus Flycatcher	*Ficedula narcissina*	M	
Mugimaki Flycatcher	*Ficedula mugimaki*	M	
Taiga Flycatcher	*Ficedula albicilla*	A	
Little Slaty Flycatcher	*Ficedula basilanica*	E	VU
Palawan Flycatcher	*Ficedula platenae*	E	VU
Cryptic Flycatcher	*Ficedula crypta*	E	
Bundok Flycatcher	*Ficedula luzoniensis*	E	
Furtive Flycatcher	*Ficedula disposita*	E	NT
Little Pied Flycatcher	*Ficedula westermanni*	R	
Daurian Redstart	*Phoenicurus auroreus*	A	
Luzon Water Redstart	*Phoenicurus bicolor*	E	VU
Blue Rock Thrush	*Monticola solitarius*	R,M	
Stejneger's Stonechat	*Saxicola stejnegeri*	A	
Pied Bush Chat	*Saxicola caprata*	R	
Northern Wheatear	*Oenanthe oenanthe*	A	
Leafbirds	**Chloropseidae**		
Philippine Leafbird	*Chloropsis flavipennis*	E	VU
Yellow-throated Leafbird	*Chloropsis palawanensis*	E	
Flowerpeckers	**Dicaeidae**		
Olive-backed Flowerpecker	*Prionochilus olivaceus*	E	
Palawan Flowerpecker	*Prionochilus plateni*	E	
Striped Flowerpecker	*Dicaeum aeruginosum*	E	
Whiskered Flowerpecker	*Dicaeum proprium*	E	
Olive-capped Flowerpecker	*Dicaeum nigrilore*	E	
Flame-crowned Flowerpecker	*Dicaeum anthonyi*	E	NT
Bicolored Flowerpecker	*Dicaeum bicolor*	E	
Red-keeled Flowerpecker	*Dicaeum australe*	E	
Black-belted Flowerpecker	*Dicaeum haematostictum*	E	VU
Scarlet-collared Flowerpecker	*Dicaeum retrocinctum*	E	VU
Cebu Flowerpecker	*Dicaeum quadricolor*	E	CR
Orange-bellied Flowerpecker	*Dicaeum trigonostigma*	R	
Buzzing Flowerpecker	*Dicaeum hypoleucum*	E	
Pygmy Flowerpecker	*Dicaeum pygmaeum*	E	
Fire-breasted Flowerpecker	*Dicaeum ignipectus*	R	
Sunbirds	**Nectariniidae**		
Brown-throated Sunbird	*Anthreptes malacensis*	R	
Grey-throated Sunbird	*Anthreptes griseigularis*	E	
Purple-throated Sunbird	*Leptocoma sperata*	NE	
Copper-throated Sunbird	*Leptocoma calcostetha*	R	

Olive-backed Sunbird	*Cinnyris jugularis*	R	
Grey-hooded Sunbird	*Aethopyga primigenia*	E	NT
Apo Sunbird	*Aethopyga boltoni*	E	NT
Lina's Sunbird	*Aethopyga linaraborae*	E	NT
Flaming Sunbird	*Aethopyga flagrans*	E	
Maroon-naped Sunbird	*Aethopyga guimarasensis*	E	
Metallic-winged Sunbird	*Aethopyga pulcherrima*	E	
Luzon Sunbird	*Aethopyga jefferyi*	E	
Bohol Sunbird	*Aethopyga decorosa*	E	
Lovely Sunbird	*Aethopyga shelleyi*	E	
Handsome Sunbird	*Aethopyga bella*	E	
Magnificent Sunbird	*Aethopyga magnifica*	E	
Orange-tufted Spiderhunter	*Arachnothera flammifera*	E	
Pale Spiderhunter	*Arachnothera dilutior*	E	
Naked-faced Spiderhunter	*Arachnothera clarae*	E	
Old World Sparrows	**Passeridae**		
Cinnamon Ibon	*Hypocryptadius cinnamomeus*	E	
Eurasian Tree Sparrow	*Passer montanus*	I	
Waxbills, Munias and allies	**Estrildidae**		
Tawny-breasted Parrotfinch	*Erythrura hyperythra*	R	
Pin-tailed Parrotfinch	*Erythrura prasina*	SU	
Green-faced Parrotfinch	*Erythrura viridifacies*	E	VU
Red-eared Parrotfinch	*Erythrura coloria*	E	NT
Dusky Munia	*Lonchura fuscans*	R	
Scaly-breasted Munia	*Lonchura punctulata*	R	
White-bellied Munia	*Lonchura leucogastra*	R	
Chestnut Munia	*Lonchura atricapilla*	R	
Java Sparrow	*Lonchura oryzivora*	I	EN
Wagtails and Pipits	**Motacillidae**		
Forest Wagtail	*Dendronanthus indicus*	M	
Eastern Yellow Wagtail	*Motacilla tschutschensis*	M	
Citrine Wagtail	*Motacilla citreola*	A	
Grey Wagtail	*Motacilla cinerea*	M	
White Wagtail	*Motacilla alba*	M	
Richard's Pipit	*Anthus richardi*	A	
Paddyfield Pipit	*Anthus rufulus*	R	
Olive-backed Pipit	*Anthus hodgsoni*	M	
Pechora Pipit	*Anthus gustavi*	M	
Red-throated Pipit	*Anthus cervinus*	M	
Buff-bellied Pipit	*Anthus rubescens*	A	
Finches	**Fringillidae**		
Brambling	*Fringilla montifringilla*	A	
Hawfinch	*Coccothraustes coccothraustes*	A	
Chinese Grosbeak	*Eophona migratoria*	A	
Japanese Grosbeak	*Eophona personata*	A	
White-cheeked Bullfinch	*Pyrrhula leucogenis*	E	
Common Rosefinch	*Carpodacus erythrinus*	A	
Red Crossbill	*Loxia curvirostra*	R	
Mountain Serin	*Chrysocorythus estherae*	R	NT
Eurasian Siskin	*Spinus spinus*	A	
Buntings	**Emberizidae**		
Chestnut-eared Bunting	*Emberiza fucata*	A	
Little Bunting	*Emberiza pusilla*	A	
Yellow-breasted Bunting	*Emberiza aureola*	A	CR
Black-headed Bunting	*Emberiza melanocephala*	A	
Yellow Bunting	*Emberiza sulphurata*	M	VU
Black-faced Bunting	*Emberiza spodocephala*	A	

SUBMISSION OF RECORDS

TAKING NOTES

Birders' observations often contribute to scientific knowledge, and there is still a lot to be learned when it comes to the Philippines' avifauna. Range distributions, migration routes, breeding habits and vocalizations are still poorly understood for many Philippine birds. New records of migrants have been documented by birders in recent years. Field notes should contain important details if they are to be acceptable for scientific records and research. Notes may contain:

1. Information on site: name of locality down to Barangay level is preferred; if GPS data can be included this is ideal.
2. Date and time of birding.
3. Environmental notes such as type of habitat, altitude, tide level and weather conditions.
4. Bird list: list of birds seen and number of birds seen; gender and age.
5. Notes of calls, behaviour, nests and other interesting or unusual observations.
6. Sketch of bird showing general shape and patterns, plumage details and field marks; especially important for hard-to-identify species and new records.
7. List of observers present and circumstances of birding activity.

Bird-trip observations and bird lists may be shared with local organizations or record committees to make sure that the trip contributes to official records, and can also help birders planning similar trips in the future.

The Wild Bird Club of the Philippines (WBCP) has been collating bird records since it started in 2005. It maintains a database of bird records and publishes an annual report summarizing the information. The WBCP Records Committee oversees and verifies the data that come from birdwatchers from all over the Philippines. A Rarities Committee has also been formed under the WBCP Records Committee, specifically to verify and study reports of rare, accidental and new country records. With the advent of technology and social media, more and more people are sharing and reporting their bird sightings. Reports and bird lists can be submitted to ebird.org. Past annual Philippine reports can be accessed on the WBCP website at birdwatch.ph.

IMPORTANT WEBSITES AND E-GROUPS

WILD BIRD CLUB OF THE PHILIPPINES

www.birdwatch.ph

The Wild Bird Club of the Philippines (WBCP) is a non-governmental organization that promotes birdwatching in the country. Comprising volunteer members, it conducts guided birdwatching trips for first-time birdwatchers.

PHILIPPINE BIRD PHOTOGRAPHY FORUM

www.birdphotoph.proboards.com
The Philippine Bird Photography Forum (PBPF) is an online forum for bird photographers to share their photos of birds photographed in the wild. It is also a resource for help in bird identification.

EBON – ELECTRONIC BIRDWATCHERS ONLINE NEWSLETTER

http://ebonph.wordpress.com
eBON is the official newsletter of the Wild Bird Club of the Philippines. It contains a variety of articles related to birdwatching.

Addresses of Key Organizations

Listed below are the government agencies and offices that assist and support birdwatching activities in the Philippines. The Department of Environment and Natural Resource (DENR) and Department of Tourism (DOT) are in charge of managing important birdwatching sites and can assist birdwatchers when they visit. Both the DENR and DOT have regional offices that can be contacted for assistance and recommendations for birdwatching activities. The Biodiversity Management Bureau (BMB) under the DENR functions as manager to the different protected areas in the country, most of which are birdwatching sites. It also serves as 'police' for any illegal activities concerning wildlife, and any such incidents may be reported to its office.

Biodiveristy Management Bureau of the Department of Environment and Natural Resources
Ninoy Aquino Parks and Wildlife Center
1100 Diliman, Quezon City, Metro Manila
+(632) 9246031 to 35
Email: bmb@bmb.gov.ph

Department of Tourism
The New DOT Building
351 Senator Gil Puyat Avenue, Makati City
+(632) 4595200 to 4595230
www.tourism.gov.ph
A list of the DOT Regional Offices can be found on its website.

There are also non-governmental organizations focused on the conservation of the environment, specifically the birdlife in the Philippines. Below are details of two of these organizations, which may also assist in birdwatching activities and in providing information on the species that they protect.

Philippine Eagle Foundation
VAL Learning Village
Ruby St., Marfori Heights, Davao City
+(63 82) 2243021
Email: info@philippineeagle.org
www.philippineeagle.org

Philippine Eagle Center
Malagos, Baguio District
Davao City

Katala Foundation
2nd Floor JMV Building, National Highway,
Bgy. Sta. Monica, Puerto Princesa City, Palawan
+(63 48) 4347693
www.philippinecockatoo.org

SUGGESTED READING

A *Guide to the Birds of the Philippines* by Robert S. Kennedy et al. is a comprehensive field guide to the birds of the Philippines. Fondly called the 'Kennedy guide' by local birdwatchers, it contains 72 full-colour plates of bird species, and accompanying text that further explains each species' description and distribution. First published in 2000, the species listed in the guide have since grown and changed, but it remains a valuable resource for anyone who wants to learn more about birds found in the Philippines.

Another reference for additional information about birds in the Philippines is BirdLife's *Philippine Red Data Book*, which lists the threatened and rare species in the country. The book provides information about the distribution and conservation status of different avifaunal species in the Philippines.

INDEX

Accipiter trivirgatus 30
virgatus 30
Acrocephalus stentoreus 113
Actenoides hombroni 78
lindsayi 78
Actitis hypoleucos 44
Aerodramus amelis 75
vanikorensis 75
Aethopyga guimarasensis 147
bella 148
flagrans 146
magnifica 149
primigenia 144
pulcherrima 147
shelleyi 148
Alauda gulgula 105
Alcedo atthis 81
Alophoixus bres 106
frater 106
Amaurornis olivacea 35
phoenicurus 35
Anas luzonica 18
Anthracoceros marchei 84
Anthus hodgsoni 155
rufulus 154
Aplonis panayensis 127
Arachnothera clarae 150
dilutior 149
Ardea cinerea 22
intermedia 23
purpurea 23
sumatrana 22
Ardeola bacchus 21
speciosa 21
Arenaria interpres 45
Artamus leucorynchus 93
Babbler, Ashy-headed 120
Chestnut-faced 121
Flame-templed 122
Golden-crowned 124
Luzon Striped 122
Melodious 121
Mindanao Pygmy 123
Visayan Pygmy 123
Balicassiao 99
Barbet, Coppersmith 86
Basilornis mirandus 128
Batrachostomus chaseni 72
septimus 71
Bee-eater, Blue-tailed 83
Blue-throated 83
Besra 30
Bittern, Black 19
Yellow 18
Bolbopsittacus lunulatus 59
Boobook, Chocolate 68
Broadbill, Visayan 90
Wattled 90
Bubo philippensis 67
Bubulcus coromandus 21
Buceros hydrocorax 84
Bulbul, Ashy-fronted 106
Grey-cheeked 106
Olive-winged 106
Palawan 106
Philippine 107
Sulphur-bellied 107
Yellowish 108
Bullfinch, White-cheeked 155
Bush-hen, Plain 35
Butorides striata 20
Buttonquail, Barred 37
Buzzard, Crested Honey 26
Cacatua haematuropygia 57
Cacomantis sepulcralis 62
Calidris canutus 46
ferruginea 47
ruficollis 46
subminuta 47
tenuirostris 45
Calliope calliope 131
Canary-flycatcher, Citrine 103
Caprimulgus affinis 74
macrurus 73
manillensis 73
Cecropis striolata 109
Centropus bengalensis 60
melanops 59
viridis 60
Ceyx argentatus 82
cyanopectus 82
melanurus 81
Chalcophaps indica 51
Charadrius dubius 39
peronii 39
Chat, Pied Bush 133
Chlidonias hybridus 49
leucopterus 49
Chroicocephalus ridibundus 48
Chrysocolaptes erythrocephalus 88
haematribon 88
lucidus 87
Cinnyris jugularis 146
Circus spilonotus 31
Cisticola, Golden-headed 115
Zitting 115
Cisticola exilis 115
juncidis 115
Cockatoo, Red-vented 57
Colasisi 57
Coleto 128
Collocalia marginata 75
Copsychus luzoniensis 132
mindanensis 131
niger 132
Coracina striata 94
Coucal, Black-faced 59
Lesser 60
Philippine 60
Crake, Red-legged 33
Slaty-legged 33
White-browed 36
Cuckoo, Philippine Drongo 62
Rusty-breasted 62
Cuckooshrike, Bar-bellied 94
Blackish 94
McGregor's 95
Culicicapa helianthea 103
Cyornis herioti 139
lemprieri 139
ruficauda 134
rufigastra 140
Dasycrotapha plateni 123
pygmaea 123
speciosa 122
superciliosus 61
Dasylophus cumingi 61
Dendrocygna arcuata 17
Dicaeum anthonyi 142
australe 142
haematostictum 142
hypoleucum 143
nigrilore 141
pygmaeum 144
trigonostigma 143
Dicrurus balicassius 99
leucophaeus 99

Dinopium everetti 87
Dove, Amethyst Brown 52
Black-chinned Fruit 55
Common Emerald 51
Cream-breasted Fruit 54
Flame-breasted Fruit 53
Red Turtle 50
Spotted 50
White-eared Brown 52
Yellow-breasted Fruit 54
Zebra 51
Dowitcher, Asian 40
Drongo, Ashy 99
Duck, Philippine 18
Wandering Whistling 17
Ducula aenea 55
bicolor 56
pickeringii 56
Eagle, Crested Serpent 27
Philippine 28
Philippine Serpent 27
Eagle-owl, Philippine 67
Edolisoma coerulescens 94
Egret, Chinese 25
Eastern Cattle 21
Little 24
Egretta eulophotes 25
garzetta 24
sacra 24
Elanus caeruleus 26
Erythropitta erythrogaster 91
kochi 91
Erythrura coloria 152
viridifacies 151
Eumyias panayensis 138
Fairy-bluebird, Philippine 126
Falco peregrinus 33
tinnunculus 32
Falcon, Peregrine 33
Falconet, Philippine 32
Fantail, Black-and-cinnamon 101
Blue-headed 100
Philippine Pied 100
Ficedula basilanica 136
crypta 137
disposita 137
platenae 136
westermanni 138
Flameback, Buff-spotted 87
Luzon 88
Red-headed 88
Spot-throated 87
Flowerpecker, Black-belted 142
Buzzing 143
Flame-crowned 142
Olive-backed 140
Olive-capped 141
Orange-bellied 143
Red-keeled 142
Palawan 141
Pygmy 144
Flycatcher, Asian Brown 135
Blue-breasted Blue 139
Cryptic 137
Furtive 137
Grey-streaked 135
Little Pied 138
Little Slaty 136
Mangrove Blue 140
Palawan 136
Palawan Blue 139
Rufous Paradise 102
Rufous-tailed Jungle 134
Turquoise 138
Frogmouth, Palawan 72
Philippine 71
Gallirallus philippensis 34
torquatus 34
Gallus gallus 16
Geokichla cinerea 130
Geopelia striata 51
Gerygone, Golden-bellied 93
Gerygone sulphurea 93
Glareola maldivarum 48
Godwit, Bar-tailed 41
Black-tailed 41
Goshawk, Crested 30
Grassbird, Striated 114
Guaiabero 59
Gull, Black-headed 48
Halcyon smyrnensis 79
Haliastur indus 31
Harpactes ardens 77
Harrier, Eastern Marsh 31
Hawk-Cuckoo, Philippine 63
Hawk-Eagle, Changeable 28
Philippine 29
Pinsker's 29
Hawk-Owl, Camiguin 71
Cebu 70
Luzon 69
Mindanao 69
Mindoro 70
Hemiprocne comata 74
Heron, Black-crowned Night 19
Chinese Pond 21
Great-billed 22
Grey 22
Javan Pond 21
Nankeen Night 20
Pacific Reef 24
Purple 23
Striated 20
Hierococcyx pectoralis 63
Himantopus himantopus 37
Hirundapus celebensis 77
giganteus 76
Hirundo rustica 108
tahitica 109
Hornbill, Luzon 85
Palawan 84
Rufous 84
Writhed 85
Horornis seebohmi 111
Hypocryptadius cinnamomeus 150
Hypothemis coelestis 102
helenae 101
Hypsipetes everetti 108
philippinus 107
Ibon, Cinnamon 150
Iole palawanensis 107
Irena cyanogastra 126
Ixobrychus flavicollis 19
sinensis 18
Junglefowl, Red 16
Kestrel, Common 32
Kingfisher, Blue-capped 78
Collared 80
Common 81
Hombron's 78
Indigo-banded 82
Philippine Dwarf 81
Southern Silvery 82
Spotted Wood 78
Stork-billed 79
White-throated 79
Winchell's 80

Kite, Black-winged 26
Brahminy 31
Knot, Great 45
Red 46
Lalage nigra 95
Lanius cristatus 97
schach 98
validirostris 98
Lark, Horsefield's Bush 105
Leptocoma calcostetha 145
sperata 145
Limnodromus semipalmatus 40
Limosa lapponica 41
limosa 41
Locustella caudata 114
Lonchura atricapilla 153
punctulata 152
Lophozosterops goodfellowi 124
Loriculus philippensis 57
Lyncornis macrotis 72
Macronus gularis 118
striaticeps 119
Magpie-Robin, Philippine 131
Malacocincla cinereiceps 120
Malacopteron palawanense 121
Malindangia mcgregori 95
Malkoha, Chestnut-breasted 61
Rough-crested 61
Scale-feathered 61
Mearnsia picina 76
Megalurus palustris 114
Megapode, Philippine 16
Megapodius cumingi 16
Merops philippinus 83
viridis 83
Microhierax erythrogenys 32
Monarch, Celestial 102
Short-crested 101
Monticola solitaries 134
Motacilla cinerea 154
tschutschensis 153
Mirafra javanica 105
Mulleripicus funebris 89
pulverulentus 89
Munia, Chestnut 153
Scaly-breasted 152
Muscicapa dauurica 135
griseisticta 135
Myna, Apo 128
Needletail, Brown-backed 76
Purple 77
Nightjar, Great Eared 72
Large-tailed 73
Philippine 73
Savanna 74
Ninox leventisi 71
mindorensis 70
philippensis 69
randi 68
rumseyi 70
spilocephala 69
Nisaetus cirrhatus 28
philippensis 29
pinskeri 29
Numenius phaeopus 42
Nuthatch, Sulphur-billed 127
Nycticorax caledonicus 20
nycticorax 19
Orthotomus chloronotus 116
derbianus 117
frontalis 116
nigriceps 118
sericeus 117
Osprey, Western 25
Otus everetti 65
fuliginosus 64
gurneyi 64
longicornis 66
mantananensis 67
megalotis 65
nigrorum 66
Owl, Eastern Grass 63
Everett's Scops 65
Giant Scops 64
Luzon Scops 66
Mantanani Scops 67
Negros Scops 66
Palawan Scops 64
Philippine Scops 65
Spotted Wood 68
Pachycephala albiventris 96
homeyari 96
philippinensis 97
Pandion haliaetus 25
Pardaliparus amabilis 104
elegans 103
Parrot, Blue-naped 58
Philippine Hanging 57
Parrotfinch, Green-faced 151
Red-eared 152
Passer montanus 151
Pelargopsis capensis 79
Penelopides manillae 85
Pernis ptilorhyncus 26
Phaenicophaeus curvirostris 61
Phapitreron amethystinus 52
leucotis 52
Pheasant, Palawan Peacock 17
Phoenicurus bicolor 133
Phyllergates cuculatus 110
heterolaemus 110
Phylloscopus borealis 111
cebuensis 112
examinandus 111
nigrorum 113
olivaceus 112
xanthodryas 111
Pigeon, Green Imperial 55
Grey Imperial 56
Philippine Green 53
Pied Imperial 56
Pipit, Olive-backed 155
Paddyfield 154
Pithecophaga jefferyi 28
Pitta sordida 92
steerii 92
Pitta, Azure-breasted 92
Hooded 92
Philippine 91
Whiskered 91
Plover, Grey 38
Little Ringed 39
Malaysian 39
Pacific Golden 38
Pluvialis fulva 38
squatarola 38
Polyplectron napoleonis 17
Porphyrio pulverulentus 36
Porzana cinerea 36
Pratincole, Oriental 48
Prioniturus luconensis 58
Prionochilus olivaceus 140
plateni 141
Psilopogon haemacephalus 86
Ptilinopus leclancheri 55
marchei 53
merrilli 54

occipitalis 54
Ptilocichla falcata 120
mindanensis 119
Pycnonotus cinereifrons 106
plumosus 106
Pyrrhula leucogenis 155
Racket-tail, Green 58
Rallina eurizonoides 33
fasciata 33
Rail, Barred 34
Buff-banded 34
Redshank, Common 42
Redstart, Luzon Water 133
Rhabdornis inornatus 129
mystacalis 129
Rhabdornis, Stripe-breasted 129
Stripe-headed 129
Rhabdotorrhinus leucocephalus 85
Rhipidura cyaniceps 100
javanica 100
nigritorquis 100
nigrocinnamomea 101
Rubythroat, Siberian
Sandpiper, Common 44
Curlew 47
Marsh 43
Wood 43
Sarcophanops samarensis 90
steerii 90
Sarcops calvus 128
Saxicola caprata 133
Scolopax bukidnonensis 40
Shama, White-browed 132
White-vented 132
Shrike, Brown 97
Long-tailed 98
Mountain 98
Sitta oenochlamys 127
Sittiparus semilavatus 104
Skylark, Oriental 105
Sparrow, Eurasian Tree 151
Spiderhunter, Naked-faced 150
Pale 149
Spilopelia chinensis 50
Spilornis cheela 27
holospilus 27
Starling, Asian Glossy 127
Sterrhoptilus dennistouni 124
Stilt, Black-winged 37
Stint, Long-toed 47
Red-necked 46
Streptopelia tranquebarica 50
Strix seloputo 68
Sunbird, Copper-throated 145
Flaming 146
Grey-hooded 144
Handsome 148
Lovely 148
Magnificent 149
Maroon-naped 147
Metallic-winged 147
Olive-backed 146
Purple-throated 145
Surniculus velutinus 62
Swallow, Barn 108
Pacific 109
Striated 109
Swamphen, Philippine 36
Swift, Philippine Spine-tailed 76
Swiftlet, Ameline 75
Grey-rumped 75
Uniform 75
Tailorbird, Black-headed 118
Grey-backed 117
Mountain 110
Rufous-fronted 116
Rufous-headed 110
Rufous-tailed 117
Trilling 116
Tanygnathus lucionensis 58
Tattler, Grey-tailed 44
Tern, Greater Crested 49
Whiskered 49
White-winged 49
Tersiphone cinnamomea 102
Thalasseus bergii 49
Thrush, Ashy 130
Blue Rock 134
Island 130
Tit-Babbler, Brown 119
Pin-striped 118
Tit, Elegant 103
Palawan 104
White-fronted 104
Todiramphus chloris 80
winchelli 80
Treeswift, Whiskered 74
Treron axillaris 53
Triller, Pied 95
Tringa brevipes 44
glareola 43
stagnatilis 43
totanus 42
Trogon, Philippine 77
Turdus poliocephalus 130
Turnix suscitator 37
Turnstone, Ruddy 45
Tyto longimembris 63
Wagtail, Eastern Yellow 153
Grey 154
Warbler, Arctic 111
Clamorous Reed 113
Japanese Leaf 111
Kamchatka Leaf 111
Lemon-throated Leaf 112
Long-tailed Bush 114
Negros Leaf 113
Philippine Bush 111
Philippine Leaf 112
Waterhen, White-breasted 35
Whimbrel, Eurasian 42
Whistler, Green-backed 96
White-vented 96
Yellow-bellied 97
White-eye, Lowland 125
Mindanao 124
Yellowish 125
Warbling 126
Woodcock, Bukidnon 40
Woodpecker, Great Slaty 89
Philippine Pygmy 86
Sooty 89
Woodswallow,
White-breasted 93
Wren-Babbler, Falcated 120
Striated 119
Yungipicus maculatus 86
Zosterops japonicus 126
meyeni 124
nigrorum 125
Zosterornis striatus 122
whiteheadi 121